The Society of Fire Protection Engineers Series

Series Editor

Chris Jelenewicz
Society of Fire Protection Engineers
Gaithersburg, MD, USA

The Society of Fire Protection Engineers Series provides rapid dissemination of the most recent and advanced work in fire protection engineering, fire science, and the social/human dimensions of fire.

The Series publishes outstanding, high-level research monographs, professional volumes, contributed collections, and textbooks.

Society for Fire Protection Engineers

Fire Exposures to Structural Elements

Springer

Society for Fire Protection Engineers
Gaithersburg, MD, USA

ISSN 2731-3638   ISSN 2731-3646 (electronic)
The Society of Fire Protection Engineers Series
ISBN 978-3-031-64027-8   ISBN 978-3-031-64025-4   (eBook)
https://doi.org/10.1007/978-3-031-64025-4

This Springer imprint is published by the registered company Springer Nature Switzerland AG
The registered company address is: Gewerbestrasse 11, 6330 Cham, Switzerland

If disposing of this product, please recycle the paper.

The SFPE Task Group on Fire Exposures to Structural Elements

Chairman

James G. Quintiere, Ph.D., FSFPE
University of Maryland

Members

Farid Alfawakhiri, Ph.D.
American Institute of Steel Construction

Andrew Buchanan, Ph.D.
University of Canterbury

Vytenis Babrauskas, Ph.D.
Fire Science & Technology Inc.

Jonathan Barnett, Ph.D., FSFPE
Worcester Polytechnic Institute

Thomas Izbicki, P.E.
Dallas Fire Department

Stephen Hill, P.E
ATF Fire Research Laboratory

Barbara Lane, Ph.D.
ARUP Fire

Sean Hunt, P.E.
Hughes Associates, Inc.

Brian Lattimer, Ph.D.
Hughes Associates, Inc.

Rodney McPhee
Canadian Wood Council

Harold Nelson, P.E., FSFPE

James Mehaffey, Ph.D
Forbitele Canada Corp.

Amai Tamim

James Milke, P.E., Ph.D., FSFPE
University of Maryland

Ian Thomas, Ph.D.
Victoria University

Christopher Wieczorek, Ph.D.
FM Global

Staff

Morgan J. Hurley, P.E.
Society of Fire Protection Engineers

Printed in the U.S.A.   Copyright ©2004 Society of Fire Protection Engineers.
All rights reserved.

Foreword

The SFPE Task Group on Fire Exposures to Structural Elements began its work in March 1998. The purpose of this guide is to provide the information and methodology needed to predict the thermal boundary condition for a fire over time. The methods contained herein are based on experimental measurements and correlations, and mostly give global rather than local results. Eventually, "CFD" methods for fire must be subjected to some of the same tests used here and judged accordingly for accuracy and application.

On September 11, 2001, the world changed, and this task took on a new life and significance. Issues identified during examination of the collapse of the World Trade Center buildings raised questions regarding the design of fire protection of structures. Indeed, the role of the fire protection engineer (FPE) in structural fire-resistance design may change and embrace more of these calculations. Presently, the architect is generally responsible for the fire protection of the structure. An engineered design method would involve:

1. A prediction of the fire over time
2. Heat transfer analysis of the structural member
3. Response of the structural system

Such full calculations will have to be dealt with by the fire protection engineer in conjunction with the structural engineer. Items 1 and 2 are more in the domain of the FPE. Note, however, that item 2 is not addressed here.

This guide was originally divided into three areas. The first included fully developed fires in compartments. Since it was an "old" area of study with many contributors, care was required to sort out the key pieces. The second area was fire plumes, or the exposure of discrete fires to elements. Since it was more recent in exposition, this work could be evaluated more easily. A third area intended for this guide included the effect of window flames on the façade and external structural elements. While this information was not included in this guide, the work of Margaret Law, "Design Guide for Fire Safety of Bare Exterior Structural Steel," *Technical Reports and Designer's Manual* (London, Ove Arup and Partners 1977), is recommended for such fire scenarios.

The work in completing this guide was mostly done voluntarily. All contributions, no matter how small, are appreciated and have enabled this guide to come to closure.

This guide is written for those with an understanding of fire and heat transfer, but should be educational and informative to a structural engineer. It includes some theoretical background for orientation, and examples to appreciate the process of calculation. It is the sixth engineering practice guide published by the Society of Fire Protection Engineers.

I take responsibility for the "theory" on compartment fires, and for the general approach of the guide. But the guide could not have been completed without the dedicated contribution of Morgan Hurley, Technical Director of SFPE. He performed the role of technical editor and personally performed the analyses and evaluations of the various methods for predicting the temperature–time curves for fully developed fires. That comparison had never been done before, and it was imperative to conduct in order to make judgment on the methods. In making those comparisons, we decided to use the CIB and Carrington data sets to serve as a benchmark. While the CIB data are of scales no more than 1.5 m in height, the Carrington tests are much more realistic in scale. However, the theory section should offset any issues of the relevance of small scale.

The section on fire plumes was developed by Brian Lattimer with the assistance of Sean Hunt. That was a significant contribution and had never been assembled before. Christopher Wieczorek organized the material describing the various approaches. Barbara Lane presented a thorough review of the time-equivalent method and drafted material on parametric equations for estimating compartment fire temperatures and durations. The time-equivalent method is limited but well known. We included this material to explicitly explain its basis and limitations.

Others made noteworthy contributions. Jonathan Barnett and his students got us started on the literature of fully developed fire, and Stephen Hill brought this to the production point in a presentation for SFPE. James Mehaffey, Ian Thomas, and Harold "Bud" Nelson were early contributors. Others, including Farid Alfawalchiri, Andrew Buchanan, Thomas Izbicki, Rodney McPhee, Amal Tamim, and James Milke, were critical readers, and Vytenis "Vyto" Babrauskas continually provided useful comments and critiques. Readers outside the Committee included Ulf Wickstrom, Takeyoshi Tanaka, Tibor Harmathy, and T.T. Lie, and for this we are greatly appreciative.

November 10, 2003   James G. Quintiere

Acknowledgment

The Society of Fire Protection Engineers wishes to acknowledge and thank the American Institute of Steel Construction, the National Fire Protection Association, the American Forest and Paper Association, and the Canadian Wood Council for their generous support of this project.

Executive Summary

Designing fire resistance on a performance basis requires three steps:

1. Estimating the fire boundary conditions
2. Determining the thermal response of the structure
3. Determining the structural response

This guide provides information relevant to estimating the fire boundary conditions resulting from a fully developed fire. Methods are provided for fully developed enclosure fires and for fire plumes. Fully developed enclosure fires can be expected in compartments with fuel uniformly distributed over their interiors. For situations where a fire would not be enclosed or for enclosures with sparse distributions or concentrated fuel packets, the methods identified in the fire plumes section should be used.

Several methods are evaluated for fully developed enclosure fires. Law's method is recommended for all roughly cubic compartments and for long, narrow compartments where $\frac{A}{A_o\sqrt{H_o}}$ does not exceed ≈ 18 m$^{-1/2}$. To ensure that predictions are sufficiently conservative in design situations, the predicted burning rate should be reduced by a factor of 1.4 and the temperature adjustment should not be reduced by Law's Ψ factor.

Law's method does not predict temperatures during the decay stage. For cases where a prediction of temperatures during the decay stage is desired, a decay rate of 7 °C/min can be used for fires with a predicted duration of 60 min or more, and a decay rate of 10 °C/min can be used for fires with a predicted duration of less than 60 min.

For long, narrow spaces in which $\frac{A}{A_o\sqrt{H_o}}$ is in the range of 45–85 m$^{-1/2}$, Magnusson and Thelandersson provide reasonable predictions of temperature and duration. For long, narrow spaces in which $\frac{A}{A_o\sqrt{H_o}}$ is approximately 345 m$^{-1/2}$, Lie's method is recommended.

For ranges of $\frac{A}{A_o\sqrt{H_o}}$ that fall outside the ranges identified above, the calculations should be performed using the methods identified for die ranges $\frac{A}{A_o\sqrt{H_o}}$ of that bound the situation of interest, and the most conservative results should be used.

For fire plumes, methods are presented for conducting a bounding analysis and for specific geometries. These geometries include flat vertical walls, comers with a ceiling, unbounded flat ceilings, and an I-beam mounted below a ceiling. Additionally, correlations are provided for axisymmetric plumes for those wishing to conduct a heat transfer analysis from first principles.

Contents

List of Figures

List of Tables

Introduction

An engineering analysis to evaluate the response of a structure during a fire must consider both the heat transfer from the fire to the structural members and the structural response of these members under the defined threat. The focus of this guide is to define the heat flux boundary condition due to the fire used in the heat transfer analysis portion of this problem. Guidance is provided for two potential fire threats: fully developed enclosure fires and local fire plumes.

In fully developed enclosure fires, the conditions (gas temperatures, velocities, and smoke levels) are assumed to be uniform throughout the entire enclosure, and all combustible contents are generally considered to be contributing to the fire size and duration. Historically, conditions inside fully developed enclosure fires have been defined by the gas temperatures inside the enclosure, and the enclosure fire section includes a review of the most widely used methods for predicting gas temperatures.

Local fire plumes may be confined to a single fuel package in intimate contact with a structural member. The thermal exposure from local fires is spatially variable and is dependent on the geometry being considered. Though local fires may not expose as large an area as enclosure fires, the heat fluxes from local fires can be considerable and should not be neglected in analysis. Heat fluxes from reasonable-size local fires can easily exceed 120 kW/m^2 and have been measured as high as 220 kW/m^2 in very large pool fires. Due to the spatial and geometric dependence, the thermal exposure from local fire plumes has historically been measured directly using heat flux gauges. Therefore, the boundary condition for local fire plumes will be provided as a measured heat flux with guidance on correcting this measurement based on the actual structural element temperature.

The methods applicable to fully developed enclosure fires should be used for compartments with fuel uniformly distributed over their interiors. For situations where a fire would not be enclosed or for enclosures with sparse distributions or concentrated fuel packets, the methods identified in the fire plumes section should be used.

Fire Exposures to Structural Elements, The Society of Fire Protection Engineers Series,
https://doi.org/10.1007/978-3-031-64025-4_1

Model Inputs

For fully developed enclosure fires, predictive methods require as input one or more of the following:

1. Fuel load
2. Dimensions of windows, doors, and other similar horizontal openings
3. Wall thermal properties

Thermal properties of walls are generally fixed very early in the design of a building. They typically do not change much during a building's lifetime. Furthermore, this is the least critical of the three variables in its effect on the fire temperature–time history. Thus, it is generally acceptable to use normal design values for the thermal properties.

Ventilation is usually handled by simply determining the potential window and door openings from the building's architectural drawings. This may not be a robust strategy since these openings may vary as a consequence of the alteration of a building. Some serious fire losses have occurred during construction or remodeling. Two examples are the One Meridian Plaza fire [1] and the Broadgate fire [2].

During construction or remodeling, the geometric aspects of a building can vary from what they are intended to be during ultimate occupancy. Uncertainty in ventilation characteristics can be addressed by a variety of techniques [3]. For example, analyses could be conducted using the range of ventilation characteristics that could reasonably be expected to occur. The ventilation characteristics that result in the most severe exposure could then be used as the basis for design. If uncertainty in ventilation characteristics is not addressed during the design, then any change that affects ventilation openings would require reanalysis to confirm that the building is still within its design basis.

Similarly, fuel loads may vary during the life of a building. During construction, periods of work may exist where the fuel load is great. Such construction fuel (and debris) may often be much greater than projected for the ultimate occupancy. Furthermore, at these times normal fire defense mechanisms—sprinklers, detectors, pull-stations, etc.—are often inoperable.

An example may be a building lobby. During normal occupancy, the expected fuel load can be trivial: perhaps a single guard's desk. Yet during construction or renovation, the lobby may hold the highest concentration of combustible building and packing materials. Another example is special events (e.g., school fair exhibits) that are sometimes staged in lobbies that are generally otherwise fuel-free.

Fuel load statistics obtained from building surveys are typically used by designers to derive their input data on fuel load. First, these statistics are "typical" values, such as 50% or 80% occurrence values. As "typical" values, these statistics would not provide bounding or conservative estimates of fire severity. Additionally, all available fuel load surveys focus solely on normal occupancy characteristics.

Methods of predicting fire exposures from fire plumes also require input values such as heat release rate or dimension of the fire source. When selecting input values for these methods, it is recommended that bounding or reasonably conservative input values be used.

Whatever input values are used, designers should clearly communicate the limits of the design to project stakeholders such as enforcement officials and building owners and operators.

Basis of Fire Resistance

Engineered fire protection design is typically performed to meet a set of goals and objectives. These goals and objectives may come from a performance-based code, from a desire to establish equivalency with a prescriptive code, or from a building owner, insurer, or other stakeholder who desires to have added safety beyond compliance with a code or standard. Fire resistance might be used as part of a strategy to achieve life safety, property protection, mission continuity, or environmental protection goals [3]. More specific objectives can be developed from these generic goals.

Structural fire resistance has historically been specified as ratings for individual structural elements based on a number of building characteristics such as occupancy type and building height. Given that the fire resistance and permissible materials of construction vary with building use and building height and area, a uniform level of performance does not result from compliance with prescriptive codes.

In the case of performance-based codes, the performance intended also may vary. The *International Code Council Performance Code* [4] states that some risk of loss of life may be acceptable, depending upon the magnitude of the event and performance group of the building. Similarly, the serviceability expected of a building varies with the event size and performance group. The National Fire Protection Association's *Building Construction and Safety Code* [5] states that structural integrity must be maintained for a sufficient time to protect occupants and enable fire fighters to perform search and rescue operations.

This guide provides a methodology to estimate the thermal aspects of a fire as they impact exposed structural members. Given those heat transfer conditions, a structural engineer can compute the effect on the structure.

Prior to designing or analyzing structural fire resistance, it is necessary to determine the objectives that the structural fire resistance is intended to meet. Guidance on determining goals and objectives can be found in the *SFPE Engineering Guide to Performance-Based Fire Protection Analysis and Design of Buildings* [3].

Accounting for Suppression

Many building codes and design guides permit a reduction in fire resistance when active fire protection systems, such as sprinklers, are used. For example, the Eurocode [6] contains an approach for accounting for interventions where the design fire load is reduced by a factor (0.0–1.0). This results in a design fire load that is less than the actual fire load.

The methods presented in this guide for predicting fire exposures are based on conditions where there is no mitigation of a fully developed fire. Analyses of fire exposures to structures in which active mitigation is considered are outside the scope of this guide.

Heat Transfer Boundary Conditions

Analyzing the thermal response of a structure requires prediction of the heat flux boundary conditions. For fire plumes, methods are provided for estimating the heat flux boundary conditions directly, although basic plume correlations are provided for those who wish to conduct a heat transfer analysis from first principles.

For enclosure fires, most of the predictive methods contained in this guide provide just the temperature boundary conditions. Determining the heat flux boundary conditions of a structure requires prediction of the gas emissivity, the absorptivity of the element, and the convective heat transfer coefficient. The absorptivity for a surface in a fully developed enclosure fire can be assumed to be 1.0 since the surface will become covered in soot. The gas emissivity will also approach 1.0 for large fires.[1] Assuming natural convection, the convective heat transfer coefficient, h_c will generally be approximately 10 W/m^{2}K, although it could be as high as 30 W/m^{2}K.[1] For conservative predictions, a convective heat transfer coefficient of 30 W/m^{2}K should be used.

For insulated materials, such as concrete or insulated steel, a bounding estimate of the heat transfer boundary condition would be to assume that the temperature of the exposed surface is equal to the surrounding gas temperature.[1]

Computer Modeling

With one exception [7], all the methods identified above for calculating the temperature–time history for a fire in a compartment are relatively simple, closed-form equations. Simple, closed-form equations are possible because of the assumptions made to solve the fundamental conservation equations, e.g., uniform conditions throughout the compartment. Indeed, even the computer model referenced above assumes a uniform temperature in the enclosure.

Many computer models exist that predict fire temperatures for user-defined heat release rates. Use of most computer fire models for predicting post-flashover fire boundary conditions requires the modeler to estimate the burning rate in the

---

[1] See the "Theory" section beginning on page 5 for a derivation of this value.

compartment using other methods. Given that the heat release rate in a post-flashover compartment fire is a function of the characteristics of the enclosure, it is difficult to apply these models without making additional simplifying assumptions. For example, by assuming that burning in the compartment is stoichiometric or ventilation-limited, a burning rate could be estimated as a constant multiplied by the ventilation characteristics of the enclosure. Pool fires could be modeled using burning rate correlations that were developed for open-air burning; however, these correlations neglect thermal feedback to the fuel from the enclosure.

Field models such as NIST's Fire Dynamics Simulator (FDS) allow abandoning the assumption that compartment gasses are well stirred [8]. Instead of modeling the enclosure as one zone, field models model an enclosure as many rectangular prisms and assume the conditions are uniform throughout each of these cells.

FDS contains pyrolysis models for solid and liquid fuels. The pyrolysis rate of the fuel is predicted by FDS as a function of the modeled heat transfer to the fuel, and thermally thick, thermally thin, and liquid fuels can be treated. Combustion is modeled by FDS using a mixture fraction model.

While FDS holds promise in calculating heat release rates in fires, it presently must be used with caution since a number of simplifications are used as a result of computational, resolution, and knowledge limitations. As stated in the FDS User's Guide, "The various phenomena [associated with modeling combustion] are still subjects of active research; thus the user ought to be aware of the potential errors introduced into the calculation" [9]. Any errors that are present with pool-like or slab-like fuels would likely be magnified when considering crib-like fuels such as furniture.

Fully Developed Enclosure Fires

Fire in enclosures may be characterized in three phases. The first phase is fire growth, when a fire grows in size and heat release rate from a small incipient fire. If there are no actions taken to suppress the fire, it will eventually grow to a maximum size, which is a function of the amount of fuel present or the amount of air available through ventilation openings. As all of the fuel is consumed, the fire will decrease in size (decay). These stages of fire development can be seen in Fig. 1.

The size (magnitude) of the fire and the relative importance of these phases (growth, fully developed, and decay) are affected by the size and shape of the enclosure; the amount, distribution, form, and type of fuel in the enclosure; the amount, distribution, and form of ventilation of the enclosure; and the form and type of materials forming the roof (or ceiling), walls, and floor of the enclosure.

The significance of each phase of an enclosure fire depends on the fire safety system component under consideration. For components such as detectors or sprinklers, the fire growth part is likely to be the most significant because it will have a great influence on the time at which they activate. The fire growth stage usually proves no threat to the structure, but if it can (e.g., if concentrated fuel packets are located close to an element), the direct heating by flames must be considered in accordance with the section on fire plumes. The threat of fire to the structure is primarily during the fully developed and decay phases [10, 11].

There are two methods of design based on fully developed compartment fires:

1. Methods that predict the boundary condition to which the structure will be exposed, from which a thermal analysis and structure analysis of the structure may be performed
2. Methods that determine an equivalent exposure to the standard temperature–time relationship

The former is the only true engineering method of designing structural fire resistance. The latter is based on determining the "equivalent" fire exposure to the "standard" temperature–time relationship, which carries an implicit assumption that the fire resistance requirements contained in prescriptive codes provide a firm design

© Society for Fire Protection Engineers 2025

7

Fire Exposures to Structural Elements, The Society of Fire Protection Engineers Series,
https://doi.org/10.1007/978-3-031-64025-4_2

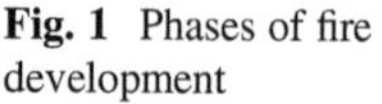

Fig. 1 Phases of fire development

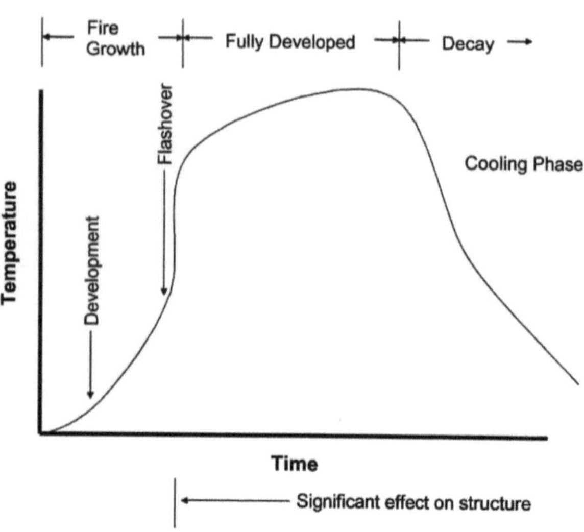

basis. While the standard temperature–time relationship provides an hourly rating, this rating is only intended to be a relative measure and does not necessarily reflect structural performance in a fire. Time-equivalent methods are further discussed only in Appendix C.

With the exception of Babrauskas' method, which allows for the consideration of pool fires, all the methods summarized in this guide have their basis in fires involving wood cribs. Although many hydrocarbon-based materials, such as plastics, have approximately twice the heat of combustion of cellulosic materials, such as wood (in other words, burning 1 kg of plastic can liberate twice the energy as burning an equal mass of wood), the use of the methods contained in this guide should be reasonable for most design scenarios.

This statement is made for two reasons. First, while real fuels are not wood cribs, cribs might approximate structural wood furniture such as desks and chairs. Other furnishings are mostly composed of large flat surfaces that would more easily vaporize fuel in a fire. These flat surfaces might be classified as "pools" since they represent a surface fully exposed to the fire. On the other hand, cribs bum from within and feel very little of the surrounding heat of the fire. The heat flux of the fire will increase vaporization over the ambient level. This depends on the fuel's heat of gasification (typically $L = 0.5$ to 1 kJ/g for liquids, 2–3 for non-charring solids, and 5–10 for charring solids).

Since the fuel volatilization rate is the heat transfer to the fuel divided by the heat of gasification of the fuel [12] and woods tend to have higher heats of gasification, wood cribs will tend to result in fires of longer duration than other fuels. In ventilation-limited fires involving non-charring fuels, the rate of airflow into the enclosure will govern the heat release rate into the enclosure, and fuels that cannot burn inside the enclosure will burn outside once they encounter fresh air.

Secondly, the primary fuel in many design or analysis situations is typically cellulosic in nature (wood, paper, etc.). While many compartments contain other fuels, the total mass of non-cellulosic fuels could be a small fraction of the mass of cellulosic materials. Design or analysis situations in which the fuels are not predominantly cellulosic and the burning is not expected to be ventilation-limited may require special attention.

Additionally, each of the methods presented in this guide is subject to the following limitations:

1. The methods are only applicable to compartments with fuel uniformly distributed over their interior. (Sparse distributions or concentrated fuel packets should be considered using the methods identified in the fire plumes section.)
2. The methods presented in this guide are only applicable to compartments having vents in walls. (Ceiling and floor vents require a special formulation, as would underground compartments having only roof vents.)
3. Only natural ventilation is considered as would occur through the wall vents. (The effect of forced ventilation and wind and stack-effect flow in tall buildings are not included.)
4. Large fires are considered whose heating effects are felt uniformly through the compartment.

Concern has been expressed that fires in long, narrow enclosures exhibit different burning behavior than fires in other types of enclosures [13], and, hence, predictive methods that were developed based on fires in compartments that are not long and narrow may not accurately predict burning behavior in long, narrow enclosures. Specifically, these long, narrow compartments with a uniformly distributed fuel load can exhibit nonuniform heating in ventilation-limited fires. To address this concern, the methods presented in this guide have been evaluated using data from fires in long, narrow enclosures in addition to compartments in which the ratio of length to width is nearly one.

Theory

It would appear that geographical reasons explain the proliferation of many models for fire resistance. Most of the work on fire resistance took place before 1970, when communication and dissemination of research in fire was limited. This might explain the existence of the different models. However, their differences are superficial for the most part, clouded by notation or parameters that might appear as different. For that reason, it was felt important to develop a theoretical base for the models. So, doing might appear to be establishing yet another model. Indeed, the contrary is intended. The purpose of this theoretical exposition is to present a rationale for the physics of the models and to show their similarities and deficiencies. It is in this context that a theoretical introduction is provided to the models that exist in the literature.

Theoretical Development

The purpose of this theoretical development is to:

1. Present the governing equations.
2. Explain and justify typical approximations.
3. Present the equations in dimensionless terms to show:

   (a) Their generality
   (b) Independence of scale
   (c) Relationship to variables used in the established methods

The common objective of all the models has been to predict the following:

1. Compartment gas temperature
2. Burning rate of the fire
3. Duration of the fire

The purpose of the studies considered has been to predict the thermal effects of fully developed building fires so that their impact on the structural members could be assessed. Fully developed fires with considerable fuel will tend to produce a fairly uniform temperature smoke layer that will descend to the floor. This will particularly occur for a large fire and relatively small vents. The radiation effects of such a fire will further tend to cause uniform heating of the contents. Consequently, the model for the fully developed fire has been an enclosure with uniform smoke or gas properties. The bounding wall surfaces are also considered uniform. The structural elements absorb a small amount of heat relative to heat loss into the wall or ceiling surfaces together with the energy loss out of the vents. These vents include the windows broken by the thermal stress of the impinging flames and heat. The model is depicted in Fig. 2.

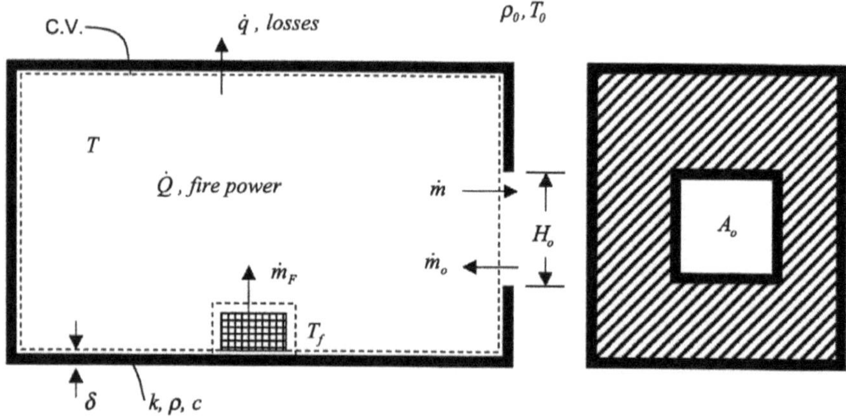

Fig. 2 Model for the fully developed fire

The conservation of mass and energy for the control volume (CV), which follows, also applies.

Mass:

$$\frac{dm}{dt} = \dot{m}_o + \dot{m}_F - \dot{m} \tag{1}$$

Energy:

$$c_v \frac{d(mT)}{dt} = \frac{c_v V}{R}\frac{dp}{dt} = \dot{Q} - \dot{q} + \dot{m}_o c_p T_F - \dot{m} c_p T \tag{2}$$

The Equation of State:

$$pV = mRT \tag{3}$$

The volume, V, is constant. The pressure, p, is nearly constant and at the ambient condition for vents that are even very small, e.g., those in the leakage category. Only for abrupt changes in the fire will pressure pulses above or below ambient occur.

The temperature slowly varies during the fully developed fire state. As a consequence, steady-state conditions can be justified.

$$\dot{m} = \dot{m}_o + \dot{m}_F \tag{4}$$

The mass flow rate from the vent ($\dot{m}$) equals the air supply ($\dot{m}_o$) and the fuel gases produced ($\dot{m}_F$). The energy equation can be written as

$$\boxed{\dot{Q} = \dot{m}c_p(T - T_o) - \dot{m}_F c_p(T_F - T_o) - \dot{q}} \tag{5a}$$

The heat losses ($\dot{q}$) consist of the heat transfer into the boundary surfaces and the radiation loss out of the vent. Some simplification can be made since $\dot{m}_F \ll \dot{m}$ and $T_F \ll T$, so that the second term on the right may be neglected.

$$\boxed{\dot{Q} = \dot{m}c_p(T - T_o) - \dot{q}} \tag{5b}$$

Wall Heat Transfer

The heat transfer into the boundary surface is by convection and radiation from the enclosure then conduction through the walls. The boundary element will be represented as a uniform material of properties:

Fig. 3 Wall heat transfer

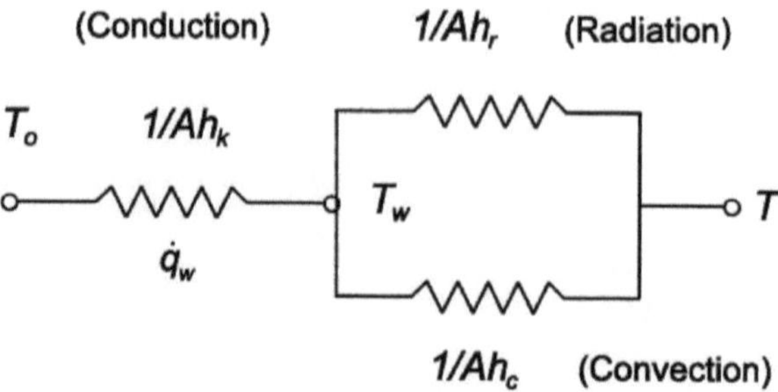

- Thickness, δ
- Thermal conductivity, k
- Specific heat, c
- Density, ρ

It conducts to a sink at T_o.

The heat transfer can be represented as an equivalent electric circuit as shown in Fig. 3.

The conductances, h_i, can be computed as follows from standard heat transfer estimates:

Convection

Convection can be estimated from natural convection [14].

$$\mathrm{Nu} = \frac{h_c H}{k} = 0.13\left[\left(\frac{g(T-T_0)H^3}{Tv^2}\right)\mathrm{Pr}\right]^{1/3} \tag{6a}$$

It gives h_c of about 10 W/m^{2}K. Under some other flow conditions, it is possible h_c might be as high as 30 W/m^{2}IC.

Conduction

Conduction might be represented as steady or unsteady. The latter is more likely. Only a finite difference numerical solution can give exact results. Most often, the following approximate analysis is used for the unsteady case assuming a semi-infinite wall under a constant heat flux. The exact solution for constant heat flux gives:

$$\dot{q}_w = A\sqrt{\frac{\pi}{4}\frac{k\rho c}{t}}(T_w - T_0) \tag{6b}$$

Table 1 Estimates of conduction for common materials

	Approximate properties		
	Concrete/Brick	Gypsum	Mineral wool
k (W/mK)	1	0.5	0.05
$k\rho c$ (W^{2}s/m^{4}K^2)	10^6	10^5	10^3
$k/\rho c$ (m^2/s)	5×10^{-7}	4×10^{-7}	5×10^{-7}

or

$$h_k = \sqrt{\frac{\pi}{4} \frac{k\rho c}{t}} \tag{6c}$$

This result for h_k can be used as an approximation for variable heat flux. For steady conduction, the exact result is

$$h_k = \frac{k}{\delta} \tag{6d}$$

The steady-state result would be considered to hold for [14]

$$t > \frac{\delta^2}{4\left(\frac{k}{\rho c}\right)}$$

Some estimations for common materials are given in Table 1. For a wall 6″ thick, $\delta \approx 0.15$ m, then

$$t = \frac{\delta^2}{4\left(\frac{k}{\rho c}\right)} = 3.1 \text{ hours}$$

Hence, most boundaries might be approximated as thermally thick since most fires would have a duration of less than 3 h.

The thermally thick case will predominate under most fire and construction conditions:

$$h_k = \sqrt{\frac{\pi}{4} \frac{k\rho c}{t}}$$

Based on $k\rho c$ of 10^3 to 10^6, it is estimated:

t (min)	h_k (W/m^{2}k)
10	0.8–26
30	0.3–10
120	0.2–5

Radiation

Radiation heat transfer can be derived from the method presented in Karlsson and Quintiere [15] (p. 170) for enclosures. It can be shown as [14]

$$\dot{q}_r = \frac{A\sigma\left(T^4 = T_w^4\right)}{\frac{1}{\varepsilon} + \frac{1}{\varepsilon_w} - 1}$$

where:

ε = Emissivity of the enclosure gas (flames and smoke)
ε_w = Emissivity of the boundary surface

Since the boundary surface will become soot-covered in a fully developed fire, $\varepsilon_w = 1$,
the gas emissivity can be represented as

$$\varepsilon = 1 - e^{-\kappa H} \tag{8}$$

where:

H = A characteristic dimension of the enclosure, its height

The absorption coefficient κ, can range from about 0.4 to 1.2 m^{-1} for typical flames (see Karlsson and Quintiere [15], p. 167). Experimental fires might use $H \approx 1$ m, while buildings generally have $H \approx 3$ m. For the smoke conditions in fully developed fires, $\kappa = 1$ m^{-1} is reasonable in the least. Hence, ε ranges from about 0.6 for a small experimental enclosure to 0.95 for realistic fires.
It follows that:

$$h_r = \varepsilon A\sigma\left(T^2 + T_w^2\right)(T + T_w) \tag{9}$$

where ε is generally nearly 1. It can be estimated for $\varepsilon = 1$, and $T = T_w$, that

$$h_r = 104 - 725 \text{ W/m}^2\text{K}$$

for $T = 500$ to 1200 °C.
From the circuit in Fig. 3, the equivalent conductance, h, allows

$$\dot{q}_w = hA(T - T_0) \tag{10a}$$

where:

$$\frac{1}{h} = \frac{1}{h_o + h_r} + \frac{1}{h_k} \tag{10b}$$

It follows from the estimates that $h \approx h_k$, which implies $T_w \approx T$ for hilly developed fires. This result applies to structural elements that are insulated, including unprotected concrete elements. Hence, predicting the fire temperature provides a simple boundary condition for the corresponding computation for the structural element. Its surface temperature can be taken as the fire temperature.

This result is very important and helps to explain why most of the methods only present the fire temperature without any detailed consideration of the heat transfer in representing the fully developed fire. From the estimates made here, the gas phase radiation and convection heat transfer have negligible thermal resistance compared to conduction into the boundary. As a consequence, the fire temperature is approximately the surface temperature. This boundary condition is "conservative" in that it gives the maximum possible heat transfer from the fire.

Radiation Loss from the Vent

From Karlson and Quintiere [15] (p. 170), an analysis of an enclosure with blackbody surfaces ($\varepsilon_w = 1$) gives the radiation heat transfer rate out of the vent of area A_o as:

$$\dot{q}_r = A_o \varepsilon \sigma (T^4 - T_0^4) + A_o(1 - \varepsilon)\sigma(T_w^4 - T_0^4) \tag{11}$$

Since ε is also near 1 and $T_w \approx T$, it follows that:

$$\dot{q}_r = A_o \sigma (T^4 - T_0^4) \tag{12}$$

This blackbody behavior for the vents has been verified [16].
The total heat losses can be written as:

$$\boxed{\begin{aligned} \dot{q} &= \dot{q}_w + \dot{q}_r \\ \dot{q} &= h_k A(T - T_o) + A_o \sigma (T^2 - T_0^2)(T + T_o)(T - T_o) \end{aligned}} \tag{13}$$

Vent Mass Flow Rate Air

The mass flow rate of air can be approximated for small ventilation as (Karlsson & Quintiere [15], p. 100)

$$\dot{m}_o = 0.5 A_o \sqrt{H_o}$$

or in general

$$\dot{m}_o = k_o \rho_o \sqrt{g} A \sqrt{H_o} \tag{14}$$

where $k_o = 0.145$ (for $\rho_0 = 1.1$ kg/m^3). This result is prevalent in all analyses, and the parameter $(A_o \sqrt{H_o})$ shows up in many experimental correlations.

The Fire–Firepower and Burning Rate

To complete the energy equation in order to solve for the temperature, the fire must be described. The heat of the flames and smoke causes the fuel to vaporize, supplying a mass flow rate, $\dot{m}_F$. While all the fuel may eventually bum, it may not necessarily burn completely in the compartment. This depends on the air supply rate. Either all the fuel is burned, or all the oxygen in the incoming air is burned. What bums inside gives the firepower within the enclosure.

Thus,

$$\begin{aligned} \dot{Q} &= \dot{m}_F \Delta H_c \phi < 1 \\ \dot{Q} &= \dot{m}_o \Delta H_{\text{air}} \phi \geq 1 \end{aligned} \tag{15}$$

The equivalence ratio, ϕ, determines if the combustion is fuel-lean (<1), or fuel-rich (>1).

$$\phi = \frac{s \dot{m}_F}{\dot{m}_o} \tag{16}$$

where:

S = Stoichiometric air-to-fuel ratio

ΔH_c = Heat of combustion (chemical heats of combustion according to Tewarson [17])

ΔH_{air} = Heat of combustion per unit mass of air $\approx$ 3kJ/g, which holds for most fuels

Note:

$$s = \frac{\Delta H_c}{\Delta H_{\text{air}}} \tag{17}$$

The mass supply rate of the fuel, $\dot{m}_F$, depends on the fuel properties, its configuration, and the heat transfer. Most studies have been done using wood cribs. These are composed of ordered layers of square sticks of side b. Gross [18] and Heskestad [19] have developed correlations to describe how they bum. For cribs that have sufficient air supply, their burning rate per unit area $\dot{m}_F'' = \dot{m}_F / A_F$ is found as:

$$\left(\dot{m}_F'' \right)_{\text{woodstick}} = C b^{-\frac{1}{2}} \left(g/m^2 s \right) \tag{18}$$

where C depends on the wood (approximately 1 mg/cms [15]).

For a range of crib experiments in compartments, Harmathy [20] gives

$$\left(\dot{m}_F''\right)_{\text{woodcrib}} = 6.2 \text{g}/\text{m}^2\text{s}$$

while Tewarson [17] gives 11 g/m^{2}s. These values give an approximation for wood, but it should be noted that, in general, it depends on the stick size.

Real fuels are not wood cribs, although cribs might approximate structural wood furniture such as desks and chairs. Other furnishings are mostly composed of large flat surfaces that would more easily vaporize fuel in a fire. These flat surfaces might be classified as "pools" since they represent a surface fully exposed to fire. On the other hand, cribs burn from within and feel very little of the surrounding heat of the fire.

In general, the mass flux of fuel produced in a fire can be represented as:

$$\dot{m}_F'' = \dot{m}_{F,o}'' f\left(Y_{o_2}\right) + \dot{q}_{\text{fire}}''/L \tag{19}$$

The fire "free"-burning flux $\dot{m}_{F,o}''$ is how the fuel would burn in ambient air. In a fire, this would be modified by the oxygen concentration the fuel experiences. Also, the heat flux of the fire $\dot{q}_{\text{fire}}''$ will increase vaporization over the ambient level. This depends on the fuel's heat of gasification (typically $L = 0.5$ to 1 kJ/g for liquids, 2 to 3 for non-charring solids, and 5 to 10 for charring solids). It is known that large fires, burning in the air, reach an asymptotic burning flux $\dot{m}_{F,\infty}''$ as their flames reach an emissivity of 1. Such values are tabulated (see Tewarson [17] or Babrauskas [21]). Since the radiant heat transfer dominates, the fuel mass loss rate in typical building compartments, where the fire is large, can be approximated as:

$$\boxed{\begin{aligned} \dot{m}_F &= \dot{m}_{F,\infty}'' A_F, \phi < 1 \\ \dot{m}_F &= \frac{\dot{m}_o}{s} = \frac{F A_F \sigma \left(T^4 - T_0^4\right)}{L}, \phi \geq 1 \end{aligned}} \tag{20}$$

Here, it is assumed that for $\phi < 1$, the **"fuel-controlled"** fire, the fire bums as a large fire with sufficient air. Such "large" fires need only achieve a burning diameter of greater than about 1 to 2 m. In the "ventilation-controlled" fire, $\phi > 1$, the fuel mass loss rate is composed of all that bums inside with the available airflow plus what is vaporized by radiant heating. The radiation geometric view factor F is, in the limits, 0 and 1, respectively, for crib-like and pool-like fuels. This expression is the governing equation for the mass loss rate. Together with the energy equation, there are two equations and two unknowns: T and $\dot{m}_F$

Development of a Solution and Dimensionless Groups

The equations will be examined to achieve insight into the form of a solution. They are not difficult to solve by iteration using a computer. However, analytical approximations can be of value. A dimensionless form of the equations will be presented to demonstrate the important variables. These variables will be used to explain the theoretical and experimental results presented in this guide in terms of the methods available in the literature.

Compartment Temperature

Substituting for the heat loss rate from Eq. 13 into the energy equation (5b) yields:

$$T - T_o = \frac{\dot{Q}}{c_p(\dot{m}_o + \dot{m}_F) + h_k A + A_o \sigma (T^2 - T_o^2)(T + T_o)} \tag{21a}$$

Dividing the numerator and denominator by $c_p \dot{m}_o T_o$ and representing

$$f(T) = (T^2 - T_o^2)(T + T_o)$$

gives

$$\frac{T - T_0}{T_0} = \frac{\dot{Q}\big/\left(c_p \dot{m}_o T_0\right)}{1 + \dfrac{\dot{m}_F}{\dot{m}_o} + \dfrac{h_k A}{c_p \dot{m}_o} + \dfrac{A_o \sigma f\left(T/T_0\right) T_0^3}{c_p \dot{m}_o}} \tag{21b}$$

By substituting for $\dot{m}_o = k_o \rho_o \sqrt{g} A_o \sqrt{H_o}$, the following dimensionless groups emerge. The dimensionless variables are presented in terms of a frequently used Q^* factor (Fig. 4).

$$\frac{\text{Chemical energy}}{\text{Vent flow energy}} = \frac{\dot{Q}}{c_p \dot{m}_o T_o} = \frac{1}{k_o} \frac{\dot{Q}}{\rho_o c_o T_o \sqrt{g} A_o \sqrt{H_o}} = \frac{Q^*}{k_o} \tag{22}$$

$$\frac{\dot{m}_F}{\dot{m}_o} = \frac{\phi}{s} \tag{23}$$

$$\frac{\text{Wall heat loss}}{\text{Vent flow energy}} = \frac{h_k A}{c_p \dot{m}_o} = \frac{1}{k_o} \frac{h_k A}{\rho_o c_p \sqrt{g} A_o \sqrt{H_o}} = \frac{Q_w^{\,*}}{k_o} \tag{24}$$

$$\frac{\text{Radiation vent loss}}{\text{Vent flow energy}} = \frac{A_o \sigma T_o^3}{c_p \dot{m}_o} = \frac{1}{k_o} \left(\frac{\sigma T_o^3}{\rho_o c_p \sqrt{g} \sqrt{H_o}}\right) = \frac{Q_r^{\,*}}{k_o} \tag{25}$$

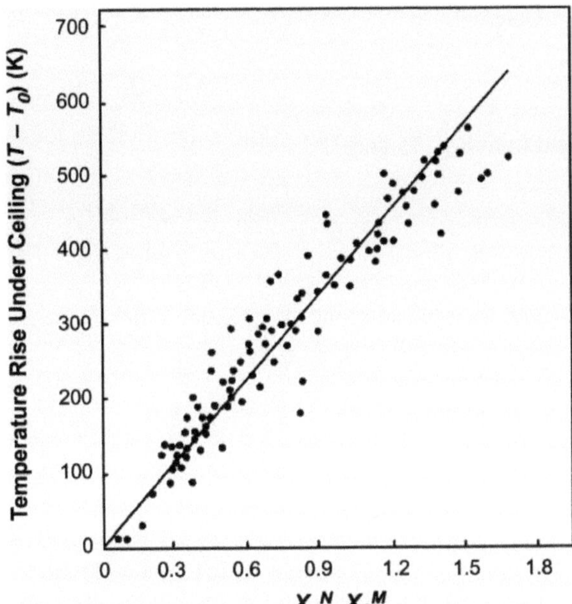

Fig. 4 MQH correlation for fuel-controlled fires. $X_1 = Q^*$, $X_2 = Q_w^*$

$$\therefore \frac{T - T_o}{T_o} = \frac{Q^*}{k_o\left(1 + \phi/_s\right) + Q_w^* + Q_r^* f\left(T/T_o\right)} \tag{26a}$$

or

$$\frac{T - T_o}{T_o} = \frac{Q^*}{k_o\left(\dfrac{\dot{m}_F}{\rho_o\sqrt{g}A_o\sqrt{H_o}}\right) + Q_w^* + Q_r^* f\left(T/T_o\right).} \tag{26b}$$

The correlation by McCaffrey, Quintiere, and Harkleroad (MQH) [22] is

$$\frac{T - T_o}{T_o} = 1.63 Q^{*2/3} Q_w^{*-1/3} \tag{27}$$

This result has only been developed from data where $\phi < 1$. However, Tanaka, Sato, and Wakamatsu [23] have applied it for $\phi > 1$ with some success.

Maximum Possible Temperature

Examine the limit of the stoichiometric adiabatic state that would yield the maximum temperature.

Here,

$$Q_w{}^* = Q_r{}^* = 0$$

and from Eqs. 15 and 22

$$Q^* = \frac{k_o \rho_o \sqrt{g} A_o \sqrt{H_o} \Delta H_{air}}{\rho_o c_o T_o \sqrt{g} A_o \sqrt{H_o}} = k_o \frac{\Delta H_{air}}{c_p T_o}$$

With $\phi = 1$, the adiabatic stoichiometric fire temperature is

$$\left(\frac{T - T_o}{T_o}\right)_{ad} = \frac{\left(\dfrac{\Delta H_{air}}{c_p T_o}\right)}{\left(1 + \dfrac{1}{s}\right)} \tag{28}$$

The experimental results for an adiabatic turbulent fire plume [24] suggest $(T - T_o)_{ad} \approx 1500\ °C$ at most. This might represent as well the maximum possible temperatures attainable in a compartment fire. The plume adiabaticity occurs due to smoke preventing the radiation loss. This occurs as the diameter of the fire becomes large. Large compartment fires can act similarly as the floor area becomes large, and only smoke is seen from the windows, particularly in an over-ventilated state, $\phi < 1$.

Burning Rate

The form of Eq. 26b suggests a corresponding dimensionless form for Eq. 20:

$$\frac{\text{Fuel burned}}{\text{Vent flow}} = \frac{\dot{m}_F}{\rho_o \sqrt{g} A_o \sqrt{H_o}} = \frac{\dot{m}''_{F,\infty} A_F}{\rho_o \sqrt{g} A_o \sqrt{H_o}}, \phi < 1$$

$$\frac{\text{Fuel burned}}{\text{Vent flow}} = \frac{\dot{m}_F}{\rho_o \sqrt{g} A_o \sqrt{H_o}} = \frac{k_o}{s} + F \frac{A_F \sigma T_o^4}{\rho_o \sqrt{g} A_o \sqrt{H_o}} \left[\left(\frac{T}{T_o}\right)^4 - 1\right], \phi \geq 1 \tag{29}$$

The last term suggests another dimensionless group governing compartment feedback.
Define

$$Q_F{}^* = \frac{\sigma T_o^4}{\rho_o \sqrt{g} L} \frac{A_F}{A_o \sqrt{H_o}} = \frac{\text{Compartment radiation to fuel}}{\text{Energy needed to vaporize fuel}} \tag{30}$$

Significant Relationships

Now, examine the values of the dimensionless variables. Estimating values are as follows:

For typical building compartments, the geometric compartment parameter is $\frac{A}{A_o\sqrt{H_o}} \approx 1$ m$^{-1/2}$ for full windows, ≈ 10 m$^{-1/2}$ for typical windows, and ≈ 100 m$^{-1/2}$ for very small vents.

Since the fuel surface area is similar and related to the room area, $\frac{A}{A_o\sqrt{H_o}}$ has a similar range.

The burning rate term can be estimated as $\frac{\dot{m}''_{F,\infty F}}{\rho_o\sqrt{g}A_o\sqrt{H_o}} \approx 10^{-3} - 1$ for wood and $\approx 10^{-2} - 10$ for liquid fuels from very large to very small vents, respectively.

The heating terms can be estimated as follows:

$Q_w^* \approx 3 \times 10^{-5} - 90$ for large to small vents, from estimates of h_k

$Q_r^* \approx 1 \times 10^{-4} - 2 \times 10^{-4}$ for $H_o \approx 3$ m

$Q_F^* \approx 1.3 \times 10^{-4} \times \dfrac{A_F}{A_o\sqrt{H_o}}$ for wood,

$1.3 \times 10^{-3} \times \dfrac{A_F}{A_o\sqrt{H_o}}$ for liquid fuels

Therefore, all terms can be significant under some circumstances.

General Form of Correlations

The dimensionless variables developed here can be used to explain the methods presented in this guide. From Eqs. 26 and 29, the approximate following solutions, in general, can be derived:

$$\frac{T - T_0}{T_0} = \text{function}\left(Q^*, \frac{\dot{m}_F}{\rho_o\sqrt{g}A_o\sqrt{H_o}}, Q_w^*, Q_r^*\right) \tag{31a}$$

$$\frac{\dot{m}_F}{\rho_o\sqrt{g}A_o\sqrt{H_o}} = \frac{\dot{m}''_{F,\infty}}{\rho_o\sqrt{g}}\frac{A_F}{A_o\sqrt{H_o}}, \phi < 1$$

$$\frac{\dot{m}_F}{\rho_o\sqrt{g}A_o\sqrt{H_o}} = \frac{k_o}{s} + F \times Q_F^*\left[\left(\frac{T}{T_o}\right)^4 - 1\right], \phi \geq 1 \tag{31b}$$

$$\therefore \frac{\dot{m}_F}{\rho_o\sqrt{g}A_o\sqrt{H_o}} = f\left(\frac{\dot{m}''_{F,\infty}}{\rho_o\sqrt{g}}\frac{A_F}{A_o\sqrt{H_o}}, \frac{k_o}{s}, Q_F^*, \frac{T}{T_o}\right)$$

$$Q^* = \frac{\dot{m}''_{F,\infty}}{\rho_o \sqrt{g}} \frac{A_F}{A_o \sqrt{H_o}} \frac{\Delta H_c}{c_p T_o}, \phi < 1$$

$$Q^* = k_o \frac{\Delta H_{air}}{c_p T_o}, \phi \geq 1 \tag{31c}$$

$$\therefore Q^* = f\left(\frac{\dot{m}''_{F,\infty}}{\rho_o \sqrt{g}} \frac{A_F}{A_o \sqrt{H_o}} \frac{\Delta H_c}{c_p T_o}\right)$$

A functional form of these equations is given from the theoretical approximation given here, but complete analytical solutions cannot be determined. Only limiting analytical solutions are possible, but these still depend on empirical factors, e.g., $\dot{m}''_{F,\infty}$, k_o, F, etc. Some limiting cases are as follows:

Large Ventilation

Large ventilation, $\frac{A}{A_o \sqrt{H_o}} < 1 \text{ m}^{-1/2}$

In this case, k_o is not a constant (Eq. 14) but depends on $\left(\frac{T}{T_o - 1}\right)^{1/2}$ due to the effect of temperature difference on the buoyancy velocity, i.e., $\dot{m}_o \sim \rho_o \nu A_o$ and $\nu \sim \sqrt{\left(\frac{T}{T_o} - 1\right) g H_o}$. For the case of large vents ($\phi < 1$), Eq. 26a can be rewritten as:

$$\frac{T - T_o}{T} \sim \frac{Q^*}{\left(\frac{T - T_o}{T}\right)^{1/2} + 10^{-3} + 10^{-5} + 10^{-4}}$$

This suggest that:

$$\boxed{\frac{T - T_o}{T} \sim Q*^{2/3} = \left(\frac{\dot{m}_F \Delta H_c}{\rho_o c_p T_o \sqrt{g} A_o \sqrt{H_o}}\right)^{2/3}} \tag{32}$$

This is consistent with the MQH correlation for $\phi < 1$ given by Eq. 27.
The mass loss rate for large ventilation ($\phi < 1$) is given directly by Eq. 31a.

$$\boxed{\frac{\dot{m}_{Fc}}{\rho_o \sqrt{g} A_o \sqrt{H_o}} = \frac{\dot{m}''_{F,\infty}}{\rho_o \sqrt{g}} \frac{A}{A_o \sqrt{H_o}} \frac{A_F}{A}} \tag{33a}$$

or alternatively

$$\boxed{\frac{\dot{m}_F}{A_F} = \dot{m}''_{F,\infty}} \tag{33b}$$

Both forms of $\dot{m}_F$ are used in the experimental correlations; however, the ratio A_F/A has not generally been included in their results. It should be recalled that, for well-ventilated wood cribs, $\dot{m}''_{F,\infty} \sim b^{-1/2}$, where b is the stick thickness.

The temperature, from Eq. 27, can be written as:

$$\frac{T - T_o}{T_o} = 1.63 \left(\frac{\dot{m}_F \Delta H_c A_F}{\rho_o c_p T_o \sqrt{g} A_o \sqrt{H_o}} \right)^{2/3} \left(\frac{h_k A}{\rho_o c_p \sqrt{g} A_o \sqrt{H_o}} \right)^{-1/3} \tag{34}$$

Small Ventilation

Small ventilation, $\frac{A}{A_o \sqrt{H_o}} \approx 100 \text{ m}^{-1/2}$

From Eq. 31b, it can be estimated for wood cribs and for large pool fires where the radiation feedback is small:

$$\boxed{\frac{\dot{m}_{FF}}{\rho_o \sqrt{g} A_o \sqrt{H_o}} \approx \frac{k_o}{s}} \tag{35}$$

The radiation feedback is negligible for cribs because of the stick blockage and for large pool fires because of obscuration by smoke. For small-scale pool fires in compartments, there can be a considerable enhancement in the burning rate due to radiation feedback.

The corresponding temperature can be estimated as follows, neglecting the vent radiation, since the vent is small.

$$\frac{T - T_o}{T_o} \approx \frac{Q^*}{k_o \left(1 + \frac{1}{s}\right)^{1/2} + Q^*_w} \tag{36}$$

However, Q^* depends on the airflow, so, by Eq. 31c,

$$\frac{T - T_o}{T_o} \approx \frac{k_o \left(\Delta H_{\text{air}} / c_p T_o \right)}{k_o \left(1 + \frac{1}{s}\right)^{1/2} + Q^*_w}$$

or

$$\boxed{\frac{T - T_o}{T_o} \approx \frac{k_o\left(\Delta H_{\text{air}}/c_p T_o\right)}{k_o\left(1 + \frac{1}{s}\right)^{1/2} + \frac{h_k A}{\rho_o c_p \sqrt{g} A_o \sqrt{H_o}}}} \tag{37}$$

For small-scale pool fires in compartments, the effect of heat feedback from the compartment is large and cannot be neglected as above.

Summary

The theory suggests that the correlations be of the following form:

- Large ventilation, $\phi < 1$ or $\dot{m}_F'' < \dfrac{k_o \rho_o \sqrt{g} A_o \sqrt{H_o}}{s A_F}$

$$\frac{T - T_o}{T_o} = 1.63 \left[\left(\frac{\dot{m}_{F,\infty}'' \Delta H_c}{c_p T_o}\right)\left(\frac{A_F}{\rho_o \sqrt{g} A_o \sqrt{H_o}}\right)\right]^{2/3} \left(\frac{h_k A}{\rho_o c_p \sqrt{g} A_o \sqrt{H_o}}\right)^{-1/3} \tag{38a}$$

$$\dot{m}_F = \dot{m}_F'' A_F \tag{38b}$$

- Small ventilation, $\phi \geq 1$ or $\dot{m}_F'' \geq \dfrac{k_o \rho_o \sqrt{g} A_o \sqrt{H_o}}{s A_F}$

$$\frac{T - T_o}{T_o} = \frac{k_o\left(\Delta H_{\text{air}}/c_p T_o\right)}{k_o\left(1 + \frac{1}{s}\right)^{1/2} + \left(\frac{h_k A}{\rho_o c_p \sqrt{g} A_o \sqrt{H_o}}\right)} \tag{39a}$$

$$\dot{m}_F = \frac{k_o}{s} \rho_o \sqrt{g} A_o \sqrt{H_o} \tag{39b}$$

Usual forms of correlations have been

$$\frac{\dot{m}_F}{A_F} \text{ vs. } \rho_o \sqrt{g} \frac{A_o \sqrt{H_o}}{A_F}$$

for wood and liquid pool fires. This would lead to results as shown in Fig. 5. A typical form for temperature is T vs. $\frac{A_o \sqrt{H_o}}{A_F}$.

Fig. 5 Approximate theoretical behavior for fuel burning rate

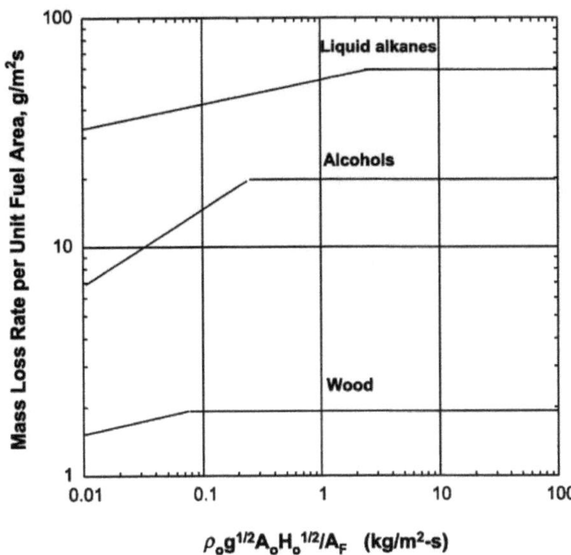

From Eq. 38a, it follows that:

$$\phi < 1, \quad \frac{T - T_o}{T_o} = 1.63 \left(\frac{\dot{m}''_{F,\infty} \Delta H_c}{c_p T_o} \right)^{2/3} \left(\frac{A_F}{A} \right)^{2/3} \left(\frac{h_k}{c_p} \right)^{-1/3} \left(\frac{A}{\rho_o \sqrt{g} A_o \sqrt{H_o}} \right)^{-1/3}$$

$$\phi \geq 1, \quad \frac{T - T_o}{T_o} = \frac{k_o \left(\Delta H_{air} / c_p T_o \right)}{k_o \left(1 + \frac{1}{s} \right)^{1/2} + \left(\frac{h_k}{c_p} \right) \left(\frac{A}{\rho_o \sqrt{g} A_o \sqrt{H_o}} \right)}$$

$$\frac{T - T_o}{T_o} \approx \frac{5 \times 10^{-3}}{0.5 + \left(\frac{h_k}{c_p} \right) \left(\frac{A}{\rho_o \sqrt{g} A_o \sqrt{H_o}} \right)}$$

This results in the following trends, as shown in Fig. 6.

In the theoretical development, the dimensionless variables that should show up in the literature correlations have been identified. The dimensionless variables contain the scaling factors that allow for the extrapolation of results over geometric scales. In addition, the dimensionless groups exhibit the proper combination of other variables including time and material properties. The theoretical results give the following functional behavior:

$$\frac{\dot{m}_F}{\rho_o \sqrt{g} A_o \sqrt{H_o}} \quad \text{and} \quad \frac{T - T_o}{T_o} = \text{Function}(Q^*, Q_w^*, Q_r^*, Q_F^*)$$

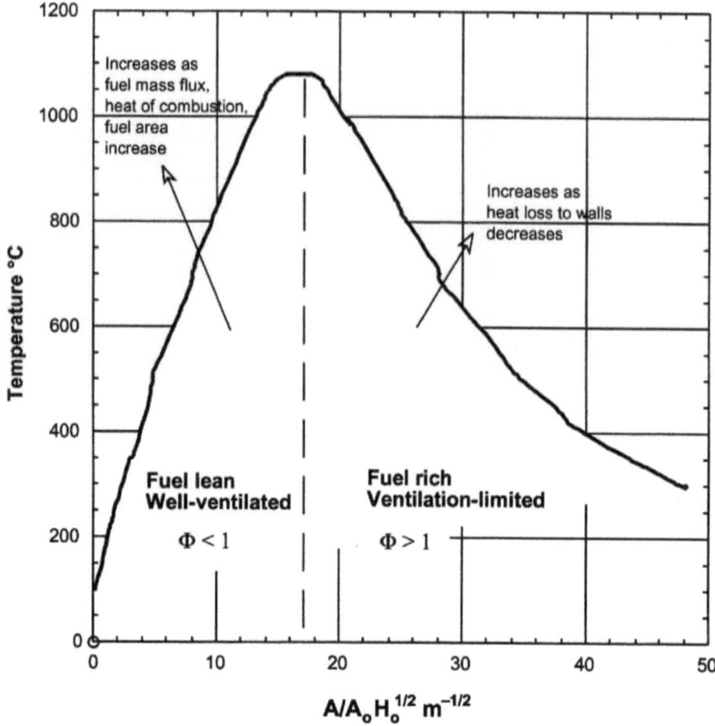

Fig. 6 Approximate theoretical behavior of compartment temperature

These dimensionless variables are not usually represented in the literature correlations in the same manner. They have equivalent surrogates.

For example:

- $\dfrac{T - T_o}{T_o}$, maximum gas temperature, is usually given as T only.

- $\dfrac{\dot{m}_F}{\rho_o\sqrt{g}A_o\sqrt{H_o}}$, burning rate/vent flow, is usually given as $\dfrac{\dot{m}_F}{A_o\sqrt{H_o}}$.

- Q^*, fire power or heat release rate; usually only ventilation-limited fire states are considered, and, consequently, this variable does not explicitly show up; however, in general,

$$Q^* = \frac{\dot{m}''_{F,\infty}}{\rho_o\sqrt{g}}\frac{A_F}{A_o\sqrt{H_o}}\frac{\Delta H}{c_pT_o}, \phi < 1$$

$$Q^* = k_o\frac{\Delta H_{\text{air}}}{c_pT_o}, \phi \geq 1$$

Note that in the latter case ($\phi > 1$) Q^* is constant. The former, or fuel-controlled, state contains the effect of fuel.

- $Q_w{}^* = \dfrac{\frac{\sqrt{k\rho c}}{t}A}{\rho_o c_p \sqrt{g} A_o \sqrt{H_o}}$, wall heat loss, is usually represented as a scaling factor for the time that allows for the temperature to be represented over dimensionless time, $t \Big/ \left(\dfrac{\rho_o \sqrt{g} A_o \sqrt{H_o}}{\sqrt{k\rho c} \quad A} \right)^2$

- $Q_r{}^* = \left(\dfrac{\sigma T_o^3}{\rho_o c_p \sqrt{g} \sqrt{H_o}} \right)$, vent radiation loss, usually does not appear in the correlations since $\sqrt{H_o}$ likely has a small variation over the range of data considered.

- $Q_F{}^* = \dfrac{\sigma T_o^4}{\rho_o c_p \sqrt{g} L} \dfrac{A_F}{A_o \sqrt{H_o}}$, enhanced fuel vaporization; for wood cribs this term is small, but for other forms of fuel in the form of flat surfaces, it can be considerable. Compared to wood cribs, it will reduce the duration of the fire, making the wood crib model conservative in design since it would give a longer duration.

Methods for Predicting Fire Exposures

Several methods are available for predicting temperatures and duration of fire exposure in a compartment. These methods are presented in an arbitrary order.

Eurocode Parametric Fire Exposure Method

The Eurocode 1, Part 2.2 [6], provides three "standard" fire curves and a parametric fire exposure. The standard fire curves include the ISO 834 curve, an external fire curve, and a hydrocarbon fire curve; these standard curves are not addressed further in this guide. The parametric fire exposure in the Eurocode was originally developed by Wickstrom [25]. Wickstrom stated [25] that this method assumes that the fire is ventilation controlled and all fuel burns within the compartment.

Wickstrom modified an approximation of the ISO 834 standard fire curve by altering the time scale based on the ventilation characteristics and enclosure thermal properties. The modified time scale compares the enclosure of interest to Magnusson and Thelandersson's "type A" enclosure with an opening factor of 0.04 m$^{1/2}$. Wickstrom found that the resulting curve approximated the ISO 834 standard fire curve.

The Eurocode states that this parametric exposure may be used for fire compartments up to 100 m^2 only, without openings in the roof, and for a maximum compartment height of 4 m. The Eurocode does not provide any basis for these limits.

The Eurocode provides the following temperature-time curve for a natural fire (also known as a parametric curve):

$$T = 1325\left(1 - 0.324e^{-0.2t^*} - 0.204e^{-1.7t^*} - 1.472e^{-19t^*}\right)$$

where:

T = Temperature (°C)
$t^* = t\Gamma$ (hours)
t = Time (hours)

$$\Gamma = \left[\frac{\left(\frac{A_o\sqrt{H_o}}{A}\right)}{0.04}\right]^2 \left(\frac{1160}{b}\right)^2$$

where the opening factor has limits of:

$$0.02 \leq \frac{A_o\sqrt{H_o}}{A} \leq 0.20 \ \left(m^{1/2}\right)$$

A_o = Area of vertical openings (m²)
H_o = Height of vertical openings (m)
A = Total area of enclosures (walls, ceilings, and floor including openings) (m²)
$b = \sqrt{k\rho c}$ (J/m² s$^{1/2}$ IC) and has the limits $1000 \leq b \leq 2000$
k = Thermal conductivity of enclosure lining (W/m-K)
ρ = Density of enclosure lining (kg/m³)
c = Specific heat of enclosure lining (J/kg-K)

For enclosures with different layers of material, $b = \sqrt{k\rho c}$ is calculated as follows:

$$b = \frac{\sqrt{\sum k_i\delta_i c_i}}{\sqrt{\sum \frac{\delta_i k_i c_i}{b_i^2}}} \ \left(J/m^2s^{1/2}K\right)$$

where:

δ_i = Thickness of layer i (m)
c_i = Specific heat of layer i (J/kg K)
k_i = Thermal conductivity of layer i (W/m K)
$b_i = \sqrt{k_i\rho_i c_i}\left(J/m^2s^{1/2}K\right)$

To account for different materials in walls, ceiling, and floor, $b = \sqrt{k\rho c}$ should be calculated as follows:

$$b = \frac{\sum b_j A_{ij}}{\sum A_{ij}}$$

where:

A_{tj} = Area of enclosure including openings with the thermal property b_j (m^2)

The temperature–time curves in the cooling phase are given by:

$$T = T_{max} - 625\left(t^* - t_d^*\right)\ (^\circ C)\ \text{for}\ t_d^* \le 0.5\ \text{hours}$$

$$T = T_{max} - 250\left(3 - t_d^*\right)\left(t^* - t_d^*\right)\ (^\circ C)\ \text{for}\ 0.5 < t_d^* < 2\ \text{hours}$$

$$T = T_{max} - 250\left(t^* - t_d^*\right)\ (^\circ C)\ \text{for}\ t_d^* \ge 2\ \text{hours}$$

where:

T_{max} = Maximum temperature (°C) in the heating phase for $t^* = t_d^*$

$$t_d^* = \frac{0.13 \times 10^{-3} q_{t,d}\Gamma A}{A_o\sqrt{H_o}}\ (\text{hours})$$

with:

$q_{t,d}$ = Design value of fuel load density related to surface area A of the enclosure, whereby $q_{t,d} = q_{f,d}A_{floor}/A$ (MJ/m^2). The limits $50 \le q_{t,d} \le 1000$ (MJ/m^2) should be observed.

$q_{f,d}$ = Design value of the fuel load density related to the surface area A_{floor} of the floor (MJ/m^2).

By making simple substitutions, t_d^* can also be expressed as:

$$t_d^* = \left(0.13 \times 10^{-3}\frac{E}{A_o\sqrt{H_o}}\right)(\text{hours})$$

where:

E = Total energy content of the fuel in the compartment, expressed by
$E = q_{f,d}A_{floor}$ or $E = q_{t,d}\,A$ (MJ)

Buchanan [10] suggested that the temperatures in the Eurocode are often too low and that it would be more accurate to scale based on a reference $\sqrt{k\rho c}$ of 1900 J/m^2 s$^{1/2}$ K. This would result in the following modified equation for Γ:

$$\Gamma_{mod} = \left[\frac{\left(\frac{A_o\sqrt{H_o}}{A}\right)}{0.04}\right]^2 \left(\frac{1900}{b}\right)^2$$

Franssen [26] noted two shortcomings of the Eurocode procedure for accounting for layers of different materials:

1. The Eurocode procedure does not distinguish which material is on the side exposed to a fire.
2. The contribution of each material to the b factor is weighted by thickness, so the adjusted b factor for an enclosure with a nominal thickness of an insulating material over a much thicker, heavier material will be biased toward the b factor of the thicker, heavier material.

Franssen therefore suggests the following alternative method of accounting for layers of different materials:

1. If a heavy material is insulated by a lighter material, the b factor for the lighter material should be used.
2. If a light material is covered by a heavier material, for example, in a sandwich panel, then a limit thickness should be calculated according to:

$$\delta_{\lim} = \sqrt{\frac{tk_1}{c_1\rho_1}}\,(\text{m})$$

where the subscript 1 indicates the properties of the material on the side exposed to the tire and t is the duration of the heating phase of the fire in seconds, which can be calculated as

$$t = \frac{t_d^*}{\Gamma} \times 3600$$

If $\delta_1 > \delta_{\lim}$, then the b factor for the heavier material should be used; otherwise,

$$b = \frac{\delta_i}{\delta_{\lim}}b_1 + \left(1 - \frac{\delta_i}{\delta_{\lim}}\right)b_2$$

Franssen observed [26] that, as the ratio between the fuel load and the ventilation factor decreases, the Eurocode predicts unrealistically short burning * durations. Therefore, Franssen suggests that if $\frac{t_d^*}{\Gamma}$ is less than 20 min, then the following procedure should be used:

1. The opening factor $\frac{A_o\sqrt{H_o}}{A}$ should be set equal to $0.13 \times 10^{-3}\frac{q_{t,d}}{0.33}$, Γ should be set equal to $\dfrac{\left(0.13 \times 10^{-3}\frac{q_{t,d}}{0.33} \times 1160\right)^2}{\left(\sqrt{k\rho c} \times 0.04\right)^2}$, and t_d^* should be set equal to $0.33 \times \dfrac{\left(0.13 \times 10^{-3}\frac{q_{t,d}}{0.33} \times 1160\right)^2}{\left(\sqrt{k\rho c} \times 0.04\right)^2}$, where 0.33 is 20 min, expressed in hours.

2. If $\frac{A_o\sqrt{H_o}}{A} > 0.04$ m$^{1/2}$ (calculated based on actual compartment geometry, not as modified above) and $q_{t,d} < 75$ MJ/m^2 $b < 1160$ J/m^2 s$^{1/2}$ K, then Γ should be set equal to $\left[1 + \frac{\left(\frac{A_o\sqrt{H_o}-0.04}{A}\right)}{0.04} \times \frac{q_{t,d}-75}{75} \times \frac{1160-\sqrt{k\rho c}}{1160}\right]$

$\frac{\left(0.13 \times 10^{-3}\frac{q_{t,d}}{0.33} \times 1160\right)}{\left(\sqrt{k\rho c} \times 0.01\right)^2}$, where $\frac{A_o\sqrt{H_o}}{A}$ is calculated based on actual compartment geometry.

Data Requirements

1. Enclosure thermal properties, k, ρ, and c. If the lining is not the same over the entire surface, the percentage of the enclosure area composed of each material is required. If multiple layers of material are present in the enclosure, the thickness of each layer is required. For thermally thick enclosure materials, it should be sufficient to account only for the innermost layer.
2. The fuel load density present in the enclosure, $q_{f,d}$.
3. The area and height of the enclosure opening(s), A_o and H_o.
4. The interior surface total area of the enclosure, including the floor and openings, A.

Data Sources

1. Thermal properties: *SFPE Handbook of Fire Protection Engineering* [27] or manufacturer's data.
2. Several surveys have been published of the mass of combustible materials per unit area for different occupancies [28–31]. Given that fire loading can vary significantly over the life of a building, uncertainty should be carefully considered. Heats of combustion are available in the *SFPE Handbook of Fire Protection Engineering* [32, 33] or other sources. To determine $q_{f,d}$, sum the products of the heat of combustion and the total mass of each material and divide this sum by the total floor surface area. Given the uncertainty that is expected in estimating the mass of materials, 40 MJ/kg is a reasonable estimate of the heat of combustion of plastics and other hydrocarbon-based materials, and 15 MJ/kg is a reasonable estimate of the heat of combustion of wood and other cellulosic materials.
3. Building characteristics can be obtained from surveys of existing buildings or architectural plans of new buildings.

Validation and Limitations

See Appendix B for comparisons of predictions with test data.

The Eurocode method, without modifications, bounds all CIB temperature data for $q_{t,d} = 50$ MJ/m^2 and most data for $q_{t,d} = 100$ MJ/m^2. The Eurocode, without modification, overpredicted the burning rate of all the CIB data and, hence, underpredicted the burning duration. In Cardington tests #1, 2, 8, and 9 $\left(\frac{A}{A_o\sqrt{H_o}} = 16 \text{ to } 18 \text{ m}^{-1/2}\right)$, the Eurocode, without modifications, bounds average temperatures, but underpredicted burning duration. In tests #3, 4, 5, and 6 $\left(\frac{A}{A_o\sqrt{H_o}} = 45 \text{ to } 345 \text{ m}^{-1/2}\right)$, the Eurocode, without modifications, reasonably predicted the burning duration but underpredicted temperature. In test #7, which was square in plan view, the Eurocode, without modification, underpredicted temperature but predicted the burning duration; however, a faster decay was predicted than was observed.

Predictions for CIB data using the Buchanan modification bound all temperature data, more so than the Eurocode method without modification, for $q_{t,d} = 50$ MJ/m^2 and $q_{t,d} = 100$ MJ/m^2. In Cardington tests #1, 2, 8, and 9 $\left(\frac{A}{A_o\sqrt{H_o}} = 16 \text{ to } 18 \text{ m}^{-1/2}\right)$, Buchanan's modification bounds peak temperature and underpredicts burning duration. In tests #3, 4, 5, and 6 $\left(\frac{A}{A_o\sqrt{H_o}} = 45 \text{ to } 345 \text{ m}^{-1/2}\right)$, Buchanan's modification reasonably predicted average temperatures and the burning duration; however, peak temperatures were underpredicted. In test #7, Buchanan's modification underpredicted temperature but predicted the duration of peak burning; however, Buchanan's modification predicted a faster decay than was observed.

The Franssen modification fell within the scatter of temperature data for values of $\frac{A}{A_o\sqrt{H_o}}$ between 0 m$^{-1/2}$ and approximately 15 m$^{-1/2}$ for $q_{t,d} = 50$ MJ/m^2 and for values of $\frac{A}{A_o\sqrt{H_o}}$ between 0 m$^{-1/2}$ and approximately 20 m$^{-1/2}$ for $q_{t,d} = 100$ MJ/m^2. For values of $\frac{A}{A_o\sqrt{H_o}}$ between 20 and 50 m$^{-1/2}$, Franssen's modification bounds all temperature data. Franssen's modification reasonably predicts peak temperatures and underpredicted the burning duration in Cardington tests #1, 2, 8, and 9 $\left(\frac{A}{A_o\sqrt{H_o}} = 16 \text{ to } 18 \text{ m}^{-1/2}\right)$. In tests #3 and 4 $\left(\frac{A}{A_o\sqrt{H_o}} = 45 \text{ m}^{-1/2}\right)$, Franssen reasonably predicts average temperatures and burning duration; however, Franssen's modification predicts a faster decay than was observed in test #4 (where the fire load was 40 kg/m^2). In tests #5 and 6 $\left(\frac{A}{A_o\sqrt{H_o}} = 85 \text{ to } 345 \text{ m}^{-1/2}\right)$ Franssen's modification slightly underpredicted average temperatures. Franssen's modification reasonably predicted burning duration in tests #5 and 6. In test #7, Franssen's modification reasonably predicted burning duration but underpredicted temperature data.

Lie's Parametric Method

Lie suggested that, if the objective is to develop a method of calculating fire resistance requirements, then it is necessary only to find a fire temperature–time curve "whose effect, with reasonable probability, will not be exceeded during the use of the building." [34] Lie developed an expression based on the series of temperature–time curves computed by Kawagoe and Selcine [35] for ventilation-controlled fires, which he proposed could be used as an approximation for the most severe fire, that is, likely to occur in a particular compartment [36].

He describes the opening factor

$$F = \frac{A_o \sqrt{H_o}}{A} \left(m^{1/2} \right)$$

where:

A_o = Area of vertical openings (m^2)
H_o = Height of vertical openings (m)
A = Total area of enclosures (walls, ceilings, and floor including openings) (m^2)

The rate of burning of the combustible materials in the enclosures is given by:

$$\dot{m}_f = 330 A_o \sqrt{H_o} (\text{kg/h})$$

where:

$$\dot{m}_f = \text{Mass burning rate of fuel}$$

Thus, if $\frac{m_f}{A}$ is the fuel load per unit area of the surfaces bounding the enclosure, the duration of the fire, τ, is

$$\tau = \frac{m_f A}{330 A A_o \sqrt{H_o}} = \frac{m_f}{330 A F}$$

where:

τ = Duration of fire (hours)

For given thermal properties of the material bounding the enclosure, the heat balance can be solved for the temperature as a function of the opening factor F. Besides depending on F, the temperature course is also a function of the thermal properties of the material bounding the enclosure.

Lie derived a series of temperature–time curves for ventilation-controlled fires in two types of enclosures: "dominantly heavy materials" and "dominantly light materials."

He found these curves could be reasonably described by the expression

$$T = 250(10F)^{0.1/F^{0.3}} e^{-F^2 t} \left[3\left(1 - e^{-0.6t}\right) - \left(1 - e^{3t}\right) + 4\left(1 - e^{-12t}\right) \right]$$

$$+ C\left(\frac{600}{F}\right)^{05} (^\circ C)$$

where:

T = Time in hours

C = Constant taking into account influence of the properties of the boundary material on the temperature:

$C = 0$ for heavy material with a density $\rho \geq 16{,}001$ cg/m^2
$C = 1$ for light materials $\rho < 1600$ kg/m^2

Lie states that the expression is valid for

$$t \leq \frac{0.08}{F} + 1 \ \text{ and } 0.01 \leq F \leq 0.15$$

If $t > (0.08/F) + 1$, a value of $t = (0.08/F) + 1$ should be used.
If $F > 0.15$, a value of $F = 0.15$ should be used.

Lie also derived an expression to define the temperature course in the decay period, over time:

$$T = -600\left(\frac{t}{\tau} - 1\right) + T_\tau (^\circ C)$$

with the condition $T = 20$ if $T < 20\ ^\circ C$
where:

T_τ = Temperature at time r ($^\circ$C)

Data Requirements

1. Enclosure density, ρ
2. The mass of fuel in the enclosure, m_f
3. The area and height of the enclosure opening(s), A_o and H_o
4. The interior surface total area of the enclosure, including the floor and openings, A

Data Sources

1. *Density: SFPE Handbook of Fire Protection Engineering* [27] *or manufacturer's data.*
2. Several surveys have been published of mass of combustible materials per unit area for different occupancies [28–31]. Given that fire loading can vary significantly over the life of a building, uncertainty should be carefully considered.
3. Building characteristics can be obtained from surveys of existing buildings or architectural plans of new buildings.

Validation and Limitations

See Appendix B for comparisons of predictions with test data.

Lie's method bounded almost all the CIB temperature data. Lie's method generally overpredicted burning rate and underpredicted burning duration for $\frac{A}{A_o\sqrt{H_o}} < 10 \text{ m}^{-1/2}$. For $\frac{A}{A_o\sqrt{H_o}} > 10 \text{ m}^{-1/2}$, predictions using Lie's method fell within the scatter of points. The data in the ventilation-controlled regime $\left(\frac{A}{A_o\sqrt{H_o}} > \sim 15 \text{ m}^{-1/2}\right)$ can be bounded by multiplying and dividing Lie's burning rate prediction by a factor of 1.8.

In Cardington tests #1, 2, 8, and 9 $\left(\frac{A}{A_o\sqrt{H_o}} = 16 \text{ to } 18 \text{ m}^{-1/2}\right)$, Lie's method predicted or slightly underpredicted average temperatures and underpredicted peak temperatures. The burning duration was underpredicted in these experiments. In test #7 $\left(\frac{A}{A_o\sqrt{H_o}} = 20 \text{ m}^{-1/2}\right)$, Lie underpredicted temperature and duration. Lie's method underpredicted temperatures in tests #3, 4, and 5 $\left(\frac{A}{A_o\sqrt{H_o}} = 45 \text{ to } 85 \text{ m}^{-1/2}\right)$; however, predictions improved as $\frac{A}{A_o\sqrt{H_o}}$ increased. Lie's method reasonably predicted the burning duration in these experiments. In test #6, $\left(\frac{A}{A_o\sqrt{H_o}} = 345 \text{ m}^{-1/2}\right)$, Lie's method reasonably predicted both temperature and duration.

Tanaka

Tanaka extended the equation for pre-flashover room fire temperature developed by McCaffrey et al. [22] to obtain equations for ventilation-controlled fire temperatures of the room of origin and the corridor connected to the room [37]. The temperature rise in a compartment can be predicted by the following equation according to McCaffrey et al.

$$\frac{\Delta T}{T_\infty}\left(\frac{T=T_\infty}{T_\infty}\right) = 1.6\left(\frac{\dot{Q}}{\sqrt{g}c_0\rho_0 T_\infty A_0\sqrt{H_0}}\right)^{2/3}\left(\frac{h_k A}{\sqrt{g}c_0\rho_0 A_0\sqrt{H_0}}\right)^{-1/3}$$

where the effective heat transfer coefficient defined as

$$h_k = \left(\frac{k\rho c}{t}\right)^{1/3} \; (\text{J/m}^2\text{sK})$$

Substituting h_k and the values of g, c_0, ρ_0, and T_∞ the equation reduces to

$$\frac{\Delta T}{T_\infty} = 0.023 \left(\frac{\dot{Q}}{A_0\sqrt{H_0}}\right)^{2/3} \left(\frac{A_0\sqrt{H_0}}{A}\right)^{1/3} \left(\frac{t}{k\rho c}\right)^{1/6}$$

where:

g = Gravity, 9.81 m/s^2
c_0 = Specific heat of air, 1.15 kJ/kg K
ρ_0 = Density of air, 1.2 kg/m^3
$\dot{Q}$ = Heat release rate (kW)
T = Temperature (K)
T_∞ = 300 K
A_0 = Area of opening (m^2)
H_0 = Height of opening (m)
A = Total surface area of room, excluding opening (m^2)
t = Time (s)
k = Thermal conductivity of enclosure lining (kW/m K)
ρ = Density of enclosure lining (kg/m^3)
c = Specific heat of enclosure lining (kJ/lcg K

Tanaka studied the effect of an opening between the corridor and the outdoors when the corridor was connected to the room of origin. His equations can be reduced where there is no opening between the room of origin and the connected corridor and can be used for predicting the temperature of a single fire room. In this case, $\dot{Q}$ becomes

$$\dot{Q} = 1500 A_0 \sqrt{H_0}$$

and substituting $\frac{\Delta T}{T_\infty} = 3.0 \left(\frac{A_0\sqrt{H_0}}{A}\right)^{1/3} \left(\frac{t}{k\rho c}\right)^{1/6}$

Tanaka's method performs all calculations in Kelvin; the equation for temperature in degrees Celsius follows.

$$T = 3.0 \left(\frac{A_0\sqrt{H_0}}{A}\right)^{1/3} \left(\frac{t}{k\rho c}\right)^{1/6} T_\infty + T_\infty - 273.15 \; (^\circ C)$$

Tanaka uses Kawagoe and Sekine's method of predicting the mass burning rate as follows:

$$\dot{m}_f = 0.1 A_o \sqrt{H_o} \; (kg/s)$$

where:

$$\dot{m}_f = \text{Mass burning rate of fuel}$$

Upon comparison of the results of the simple equations to results of a more detailed computer model, Tanaka refined the equations to improve accuracy. Tanaka defined the parameter $\beta_{F,1}$ as $\beta_{F,1} = K_F^{2/3} \cdot F_{O,F}^{1/3} \cdot \eta^{1/6}$ and the equations for temperature of the fire room are

$$\frac{\Delta T}{T_\infty} = \beta_{F,1}\left(2.50 + \beta_{F,1}\right) \; \left(\beta_{F,1} \le 1.00\right)$$

or

$$\frac{\Delta T}{T_\infty} = \beta_{F,1}\left(4.50 - \beta_{F,1}\right) \; \left(\beta_{F,1} > 1.00\right)$$

where:

$$\eta = \frac{t}{k\rho c}$$

$$F_{O,F} \frac{A_O\sqrt{H_O}}{A}$$

and K_f reduces to 1. $\beta_{F\,1}$ can be simplified to

$$\beta_{F,1} = \left(\frac{A_O\sqrt{H_O}}{A}\right)^{1/3} \left(\frac{t}{k\rho c}\right)^{1/6}$$

The equation for temperature must be re-dimensionalized and converted to degrees Celsius in the same manner as before.

Data Requirements

1. Enclosure thermal properties, k, ρ, and c
2. The height and area of the enclosure opening(s), A_o and H_o
3. The interior total surface area of the enclosure, including the floor, but excluding the opening(s), A
4. The mass of fuel in the enclosure, wy

Data Sources

1. Thermal properties: *SFPE Handbook of Fire Protection Engineering* [27] or manufacturer's data.

2. Several surveys have been published of mass of combustible materials per unit area for different occupancies [28–31]. Given that fire loading can vary significantly over the life of a building, uncertainty should be carefully considered.
3. Building characteristics can be obtained from surveys of existing buildings or architectural plans of new buildings.

Validation and Limitations

See Appendix B for comparisons of predictions with test data.

Both of Tanaka's methods bounded all the CIB temperature data; however, the refined method more closely approximates the values. Both Tanaka's simple and refined methods use the same correlation for the burning rate. Tanaka's methods overpredicted burning rate and underpredicted burning duration for $\frac{A}{A_o\sqrt{H_o}} < 10 \text{ m}^{-1/2}$. For $\frac{A}{A_o\sqrt{H_o}} > 10 \text{ m}^{-1/2}$, Tanaka's methods fell within the scatter of points. Burning rate for those tests in the ventilation-controlled regime $\left(\frac{A}{A_o\sqrt{H_o}} > \sim 15 \text{ m}^{-1/2}\right)$ bounded by multiplying Tanaka's prediction by a factor of 1.6 and dividing by a factor of 1.9.

Tanaka's simple and refined methods overpredict temperatures but underpredict duration for Cardington test #1, 2, 8, and 9 $\left(\frac{A}{A_o\sqrt{H_o}} = 16 \text{ to } 18 \text{ m}^{-1/2}\right)$. The simple method overpredicts temperature and reasonably predicts duration for test #7 $\left(\frac{A}{A_o\sqrt{H_o}} = 20 \text{ m}^{-1/2}\right)$, while the refined method reasonably predicts both values. The simple method greatly overpredicts temperature, and the refined method reasonably predicts average temperature for test #3, 4, and 5 $\left(\frac{A}{A_o\sqrt{H_o}} = 45 \text{ to } 85 \text{ m}^{-1/2}\right)$, while both underpredict duration. For test #6, Tanaka's simple method overpredicts temperature, and the refined method underpredicts temperature, yet both reasonably predict duration. The quality of temperature predictions using Tanaka's refined method decreases as $\frac{A}{A_o\sqrt{H_o}}$ increases.

Magnusson and Thelandersson Parametric Curves

Magnusson and Thelandersson [38] studied the variations in the development of energy, the effects of air supply, and the resulting evolution of gases with time in the course of a fire. They determined the temperature of the combustion gases from wood fuel fires, in an enclosed space as a function of time, under different conditions.

Magnusson and Thelandersson made adjustments to Kawagoe's work to accommodate the effect of a cooling phase since Kawagoe and Sekine's work is more applicable to the flame phase process of fire development (Fig. 7).

They used the equation of energy balance derived by Kawagoe and Seldne [35]:

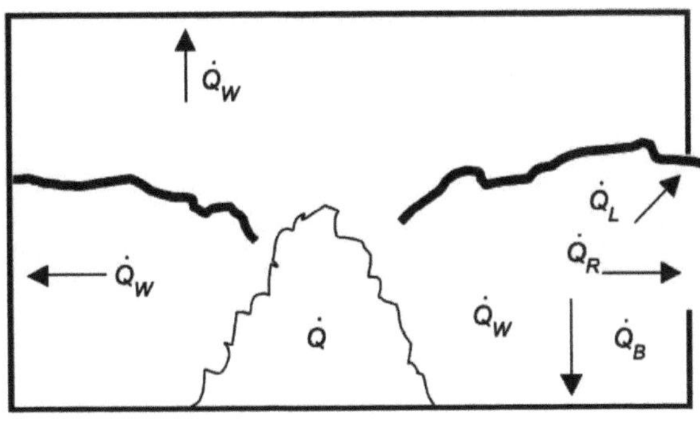

Fig. 7 Schematic illustration of the heat balance equation terms [38]

$$\dot{Q} = \dot{Q}_L + \dot{Q}_W + \dot{Q}_R + \dot{Q}_B$$

where:

$\dot{Q}$= Rate of heat energy released per unit time during combustion

$\dot{Q}_L$ = Rate of heat energy withdrawn per unit time from the enclosed space owing to replacement of hot gases by cold air

$\dot{Q}_W$ = Rate of heat energy withdrawn per unit time from enclosed space through the wall floor or ceiling and roof structures

$\dot{Q}_R$ = Rate of heat energy withdrawn per unit time from the enclosed space by radiation through the openings in the enclosed space

$\dot{Q}_B$ = Rate of the heat energy stored per unit time in the gas volume that is contained in the enclosed space

Magnusson and Thelandersson also use the opening factor, $\frac{A_o\sqrt{H_o}}{A}$ where:

A_o = Area of opening (m^2)
H_o = Height of opening (m)
A = Total surface area of room, excluding opening (m^2)

Magnusson and Thelandersson evaluated eight specific types of enclosures and developed temperature–time curves for each, assuming wood fuel. The opening factor and the fuel load were varied for each of the eight types of enclosures, and temperature as a function of time was presented in both graphic and tabular formats. Figure 8 shows examples of temperature–time curves developed by Magnusson and Thelandersson.

For practical design, they suggest that the designer choose the type of enclosed space most similar to one of the eight types with respect to the thermal properties of the bounding structure. The designer should then determine the opening factor and the fuel load for his/her case, and finally interpolate linearly, if necessary.

Alternatively, the designer can choose a curve that is determined without interpolation so as to be on the safe side; tire designer chooses the next higher value of opening factor and fuel load.

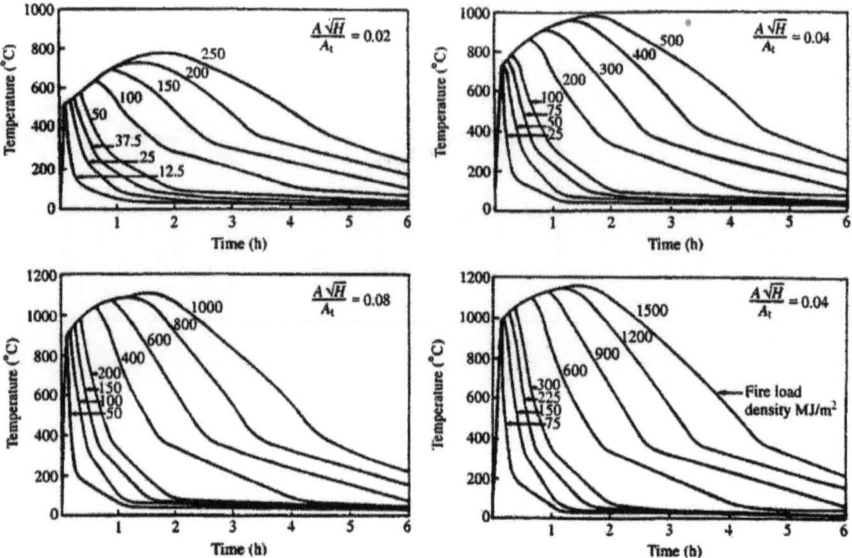

Fig. 8 Examples of temperature–time curves

Data Requirements

1. Construction materials of the enclosure
2. The fuel load density (related to the surface area of the enclosure), q
3. The area and height of the enclosure opening(s), A_o and H_o
4. The interior surface total area of the enclosure, including the floor and openings, A

Data Sources

1. Several surveys have been published of mass of combustible materials per unit area for different occupancies [28–31]. Given that fire loading can vary significantly over the life of a building, uncertainty should be carefully considered. Heats of combustion are available in the *SFPE Handbook of Fire Protection Engineering* [27, 33] and other sources. (Note that values expressed in MJ/kg must be converted to Mcal/kg by multiplying by 0.239.) To determine q, sum the products of the heat of combustion and the total mass of each material and divide this sum by the total enclosure surface area. Given the uncertainty that is expected in estimating the mass of materials, 40 MJ/kg (10 Mcal/kg) is a reasonable estimate of the heat of combustion of plastics and other hydrocarbon-based materials, and 15 MJ/kg (4 Mcal/kg) is a reasonable estimate of the heat of combustion of wood and other cellulosic materials.

2. Building characteristics can be obtained from surveys of existing buildings or architectural plans of new buildings.

Validation and Limitations

See Appendix B for comparisons of predictions with test data.

For values of $\frac{A}{A_o\sqrt{H_o}}$ for which Magnusson and Thelandersson provide predictions, Magnusson and Thelandersson's predictions bounded the temperature data from the CIB tests. Magnusson and Thelandersson's predictions overpredicted burning rate and underpredicted burning duration for $\frac{A}{A_o\sqrt{H_o}} < 13 \text{ m}^{-1/2}$. However, for $\frac{A}{A_o\sqrt{H_o}} < 13 \text{ m}^{-1/2}$, Magnusson and Thelandersson's predictions fell within the scatter of points. Those tests in the ventilation-controlled regime $\left(\frac{A}{A_o\sqrt{H_o}} > 15 \text{ m}^{-1/2}\right)$ can be bounded by multiplying Magnusson and Thelandersson's prediction by a factor of 1.3 and dividing by a factor of 2.3.

Magnusson and Thelandersson's method predicts peak temperatures, but underpredicts duration, for Cardington tests #1, 2, 8, and 9 $\left(\frac{A}{A_o\sqrt{H_o}} = 16 \text{ to } 18 \text{ m}^{-1/2}\right)$. Magnusson and Thelandersson reasonably predict average temperatures and duration for Cardington tests #3 and 4 $\left(\frac{A}{A_o\sqrt{H_o}} = 45 \text{ m}^{-1/2}\right)$. For test #5 $\left(\frac{A}{A_o\sqrt{H_o}} = 85 \text{ m}^{-1/2}\right)$, Magnusson and Thelandersson reasonably predict duration but slightly underpredict temperature. In Cardington Test #7, which was square in plan view, predictions made using Magnusson and Thelandersson's method almost coincided with the data.

Harmathy

Harmathy published a method for predicting burning rates and heat fluxes in compartment fires with cellulosic fuels [39, 40]. Harmathy's method is based on theory, with a number of simplifications and comparisons of data to define constants. The methods that Harmathy presented are applicable to fully developed fires in compartments that are ventilation-limited or fuel-bed-controlled.

Harmathy developed a method for calculating the burning rate as follows:

$$\dot{m}_f = 0.0062 A_f \quad \text{for} \frac{\rho_0\sqrt{g}A_o\sqrt{H_o}}{A_f} \geq 0.263$$

$$\dot{m}_f = 0.236\rho_0\sqrt{g}A_o\sqrt{H_o} \quad \text{for} \frac{\rho_0\sqrt{g}A_o\sqrt{H_o}}{A_f} < 0.263$$

where:

$\dot{m}_f$ = Mass burning rate of fuel (kg/s)
ρ_0 = Density of air (kg/m^3)
g = Gravitational constant (9.81 m/s^2)
A_o = Area of ventilation opening (m^2)
H_o = Height of ventilation opening (m)
A_f = Surface area of fuel (m^2)

Harmathy notes that a "critical regime" exists where the burning rate is poorly predicted using the above equations. This regime is the range

$$0.235 < \frac{\rho_0\sqrt{g}A_o\sqrt{H_o}}{A_f} < 0.290.$$

Harmathy established the duration of the fully developed burning period as the time that the combustible mass remaining in the compartment is 80% or more of the initial mass. Using this definition, Harmathy established the following expressions for the duration of the fully developed fire exposure:

$$\tau = \frac{151m_f}{A_f} \quad \text{if} \frac{\rho_0\sqrt{g}A_o\sqrt{H_o}}{A_f} \geq 0.263$$

$$\tau = 39.7\frac{m_f}{\rho_0\sqrt{g}A_o\sqrt{H_o}} \quad \text{if} \frac{\rho_0\sqrt{g}A_o\sqrt{H_o}}{A_f} < 0.263$$

where:
   τ = Time of primary (fully developed) burning (s)

Harmathy provides a method of computing the effective heat flux from the compartment fire to objects within the compartment as follows:

$$\bar{q}_E = \frac{1}{A}\left[\dot{m}_f(0.932\beta\Delta H_v + 0.068\beta\Delta H_c) - (\dot{m}_c + \dot{m}_f)c(\zeta T - T_0) - \sigma\frac{A_o}{3}(T^4 - T_0^4)\right]$$

where:

$\bar{q}_E$ = Average effective heat flux (W/m^2)
$\beta = 1$ if $l \leq H$
$\beta = \left(\frac{H}{l}\right)^{3/2}$ if $l > H$
$l = 0.75A_f^{1/3}$ if $\dfrac{\rho_0\sqrt{g}A_o\sqrt{H_o}}{A_f} \geq 0.263$
$l = 1.17\left(\rho_0\sqrt{g}A_o\sqrt{H_o}\right)^{1/3}$ if $\dfrac{\rho_0\sqrt{g}A_o\sqrt{H_o}}{A_f} < 0.263$
$\dot{m}_o = 0.145\rho_0\sqrt{g}A_o\sqrt{H_o}$
$\dot{m}_0$ = Mass flow rate of air (kg/s)
l = Flame length (m)
T = Temperature in compartment (K)

T_o = Ambient temperature (K)
ζ = Factor ($-$) (1.05)

To apply the temperature for heat flux, it is necessary to determine T. To do this, it is first necessary to determine the surface temperature of boundary elements in the compartment; Harmathy recommends the following equation to determine the surface temperature of boundary elements:

$$T_w = T_o + \frac{2\bar{q}_E}{k}\left(\frac{\kappa t}{\pi}\right)^{1/2}$$

where:

T_w = Surface temperature of boundary elements (K)
$\kappa = k/\rho c$
k = Thermal conductivity of enclosure lining (W/m-K)
ρ = Density of enclosure lining (kg/m^3)
c = Specific heat of enclosure lining (J/kg-K)
t = Time (s)

Harmathy states that, where boundary materials e not homogeneous, a weighted average can be used. Also, Harmathy suggests that, where lining materials are layered, the properties of the inner layer may be used.

$$T \approx \left\{\frac{\bar{q}_E}{\sigma\eta} + \left[T_0 + \sqrt{2}\frac{\bar{q}_E}{k}\left(\frac{\kappa t}{\pi}\right)^{1/2}\right]^4\right\}^{1/4} \quad (K)$$

where:

σ = Stefan-Boltzmann constant, 5.67×10^{-8} W/m^2 T^4
η = Factor ($-$) (0.9)
τ = Burning duration (s)

This results in two equations and two unknowns. Harmathy suggests selecting a value for T and inserting it into the equation for determining the effective heat flux. The calculated value for $\bar{q}_E$ can then be substituted into the equation for determining T, which can be substituted back into the equation for determining $\bar{q}_E$. This process of iteration can be repeated until the changes in calculated values are small.

Decay

Harmathy suggests that during the decay period the temperature can be calculated as follows:

$$\frac{T - T_0}{T(\tau) - T_0} = \exp[-0.00067(t - \tau)] \ (\mathrm{K})$$

Data Requirements

1. Enclosure thermal properties, k, ρ, and c
2. The density and specific heat of air, ρ_0 and c_0
3. The total mass of fuel, m_f
4. The total free surface area of the fuel, A_f
5. The area and height of the enclosure opening(s), A_o and H_o
6. The interior surface total area of the enclosure, including the floor but not including openings, A, and the height of the interior of the enclosure, H
7. Heat of combustion of the volatiles and char, ΔH_v and ΔH_c

Data Sources

1. Thermal properties: *SFPE Handbook of Fire Protection Engineering* [27] or manufacturer's data.
2. Density and specific heat of air: 1.2 kg/m^3 and 1150 J/kg-K, respectively.
3. The surface area-to-mass ratio of the fuel typically varies between 0.1 and 0.4 m^2/kg for larger wood cribs and conventional furniture and more often varies between 0.12 and 0.18 m^2/kg [40].
4. For wood products, the heat of combustion of volatiles can be assumed to be 16.7 × 10^6 J/kg, and the heat of combustion of char can be taken as 33.4 × 10^6 J/lcg [39].
5. Several surveys have been published of the mass of combustible materials per unit area for different occupancies [28–31]. Given that fire-loading cars vary significantly over the life of a building, uncertainty should be carefully considered.
6. Building characteristics can be obtained from surveys of existing buildings or architectural plans of new buildings.

Validation and Limitations

See Appendix B for comparisons of predictions with test data.

Due to the iterative nature of Harmathy's method, it was not possible to compare predictions to the CIB temperature data. For ventilation-limited fires, predictions made using Harmathy's method fell within the scatter of the test points. The burning rate data can be bounded by multiplying and dividing predictions made using Harmathy's method by a factor of 1.8.

In the CIB tests, for fuel-controlled fires, $\frac{\dot{m}_F}{A_o\sqrt{H_o}}$ fell within a range of approximately 0.003 $\frac{A}{A_o\sqrt{H_o}}$ to 0.012 $\frac{A}{A_o\sqrt{H_o}}$. Since $A_o\sqrt{H_o}$ occurs in the denominator of both terms, $\dot{m}_F$ ranged from approximately 0.003 to 0.012 A. In the CIB tests, the average value of A_F/A was approximately 0.75. Substituting, $\dot{m}_F$ ranged from approximately 0.002 to 0.009 (kg/m^{2}s) A_F. Therefore, multiplying Harmathy's burning rate prediction for fuel-controlled fires by 1.5 and dividing it by 2.8 bounds most of the data.

Harmathy's method underpredicted temperature and duration in Cardington tests #1, 2, 8, and 9 $\left(\frac{A}{A_o\sqrt{H_o}} = 16 \text{ to } 18 \text{ m}^{-1/2}\right)$, and in tests #3, 4, and 5 $\left(\frac{A}{A_o\sqrt{H_o}} = 45 \text{ to } 85 \text{ m}^{-1/2}\right)$ overpredicted temperatures but underpredicted duration. Harmathy reasonably predicted duration in test #6 $\left(\frac{A}{A_o\sqrt{H_o}} = 345 \text{ m}^{-1/2}\right)$ but overpredicted temperature. In test #7 $\left(\frac{A}{A_o\sqrt{H_o}} = 20 \text{ m}^{-1/2}\right)$, which was square in plan view, Harmathy's method predicted duration well but overpredicted temperature.

Babrauskas

The software program COMPF was completed and released to the public in 1975 [7]. The documentation of the program comprised a user's guide and a complete source code listing of the program. A comprehensive presentation of the theory was then presented as part of Babrauskas' Ph.D. dissertation [41]. The portions of the dissertation pertinent to COMPF theory were subsequently made available as a pair of journal articles [42].' [43]

The original COMPF program treated only wood crib fuels or else arbitrary fuels for which burning rate data were known and could be inputted. A second version, COMPF2 [44], allowed treatment of liquid and thermoplastic pools.

During the development of COMPF, it was realized that not all the input data that might be desired would necessarily be available to the designer. Thus, the idea of "pessimization" was introduced. In addition to running in a purely deterministic mode, two other modes of computation were available. In one case, the fuel mass loss rate would be computed as usual, but window ventilation would not be set to the maximum open area. Instead, the instantaneous open area was computed by the program to always be a value that would lead to the highest room temperature (up to the maximum full opening size). In a second pessimization mode, the window ventilation would have a fixed value, but the fuel mass loss rate would be instantaneously adjusted to give the highest room temperature.

Babrauskas used COMPF2 to create a series of closed-form algebraic equations that can be used to estimate temperatures resulting from fully developed fires. According to Babrauskas, estimations made using the closed-form equations are accurate to within 3% to 5% of COMPF2 predictions, typically closer to 3% [45]. The general equation follows:

$$T = T_o + (T^* - T_o) * \theta_1{}^*\theta_2{}^*\theta_3{}^*\theta_4{}^*\theta_5(^\circ C)$$

where:

T = Temperature in compartment (°C)
T_q = Ambient temperature (°C)
T^* = Constant = 1452 °C

The first variable, θ_1, known as the burning rate stoichiometry, is found for two separate regimes using:

$\theta_1 = 1.0 + 0.51\ \ln\phi$ for $\phi < 1$ for the fuel-lean regime, or
$\theta_1 = 1.0 - 0.05(\ln\phi)^{5/3}$ for $\phi > 1$ for the fuel-rich regime

$$\phi = \frac{\dot{m}_f}{\dot{m}_{st}}$$

$$\dot{m}_{st} = \frac{A_o\sqrt{H_o}}{2s}$$

where:

A_o = Area of ventilation opening (m^2)
H_o = Height of ventilation opening (m)
$\dot{m}_f$ = Mass burning rate of fuel (kg/s)
$\dot{m}_{st}$ = Mass burning rate of fuel at stoichiometry (kg/s)
ϕ = Equivalence ratio (−)
s = Ratio such that 1 kg fuel + s kg air = $(1 + s)$ kg products
ΔH_c = Heat of combustion (MJ/Kg)
σ = Stefan–Boltzmann constant (5.67×10^{-11} kW/m^{-2}-K^{-4})

For pool fires,

$$\theta_1 = 1.0 - 0.092(-\ \text{ln}\eta)^{1.25}$$

$$\eta = \frac{A_o\sqrt{H_o}}{A_f}\ \frac{0.5\Delta H_p}{s\sigma(T^4 - T_b^4)}$$

where:

T_b = Fuel boiling point (K)
A_f = Surface area of fuel (m^2)
ΔH_p = Heat of vaporization of liquid (kJ/kg)

Additionally, the heat release rate may be used in place of the mass loss rate according to the following equation:

$$\phi = \frac{\dot{Q}}{1500 A_o \sqrt{H_o}}$$

where:

$\dot{Q}$ = Heat release rate (kW)

The second variable, θ_2, accounts for wall steady-state losses and is determined using the following equation:

$$\theta_2 = 1.0 - 0.94 \ \exp\left[-54 \left(\frac{A_o \sqrt{H_o}}{A} \right)^{2/3} \left(\frac{\delta}{k} \right)^{1/3} \right]$$

where:

A = Interior surface area of the enclosure, excluding the floor and openings
δ = Thickness of wall surface (m)
k = Thermal conductivity of enclosure lining (W/m-K)

Transient wall losses are incorporated into θ_3 as follows:

$$\theta_3 = 1.0 - 0.92 \ \exp\left[-150 \left(\frac{A_o \sqrt{H_o}}{A} \right)^{3/5} \left(\frac{t}{k\rho c} \right)^{2/5} \right]$$

where:

t = Time (hours)
c = Specific heat of enclosure lining (J/kg-K)
ρ = Density of enclosure lining (kg/m^3)

If only steady-state temperatures need to be evaluated, $\theta_3 = 1$.

The variable θ_4 accounts for the effect that the height of a vent in relation to the total vent size can have on a compartment's radiative losses and is given as follows:

$$\theta_4 = 1.0 - 0.205 H_o^{-0.3}$$

The final variable, θ_5, describes the effect of combustion efficiency on the compartment temperature. This variable takes into account the fact that the gases in the compartment may not be completely mixed and is found using:

$$\theta_5 = 1.0 + 0.5 \ln b_p$$

where:

b_p = Maximum combustion efficiency (ranges from 0.5 to 0.9)

Data Requirements

1. Mass pyrolysis rate of fuel, $\dot{m}_f$, or heat release rate, $\dot{Q}$
2. The ratio s where 1 kg fuel $+ s$ kg air $= (1+s)$ kg products or the chemical formula of the fuel
3. The area and height of the enclosure opening(s), A_o and H_o
4. The interior surface total area of the enclosure, A, not including the floor or openings
5. For liquid fuels, the heat of vaporization of the liquid, ΔH_p, the fuel boiling point, T_b, and the pool area, A_f
6. Enclosure thermal properties, k, ρ, and c, and the thickness of the enclosure, δ
7. The combustion efficiency, b_p

Data Sources

1. For ventilation-controlled fires, the mass pyrolysis rate of fuel can be calculated from $\dot{m} = 0.12 A_o \sqrt{H_o}$ [44]. For fuel-controlled fires, Harmathy [39] suggests $\dot{m} = 0.0062 A_f$, where A_f is the free surface area of the fuel. The surface area-to-mass ratio of the fuel typically varies between 0.1 and 0.4 m^2/kg for larger wood cribs and conventional furniture and more often varies between 0.12 and 0.18 m^2/kg [40].
2. For hydrocarbon-based fuels, s can be calculated as follows:

$$C_xH_yO_z + wO_2 + w\left(\frac{79}{21}\right)N_2 \rightarrow xCO_2 + \frac{y}{2}H_2O + w\left(\frac{79}{21}\right)N_2$$

where

$$w = \frac{2x + y/2 - z}{2}$$

and

$$s = \frac{[w = w(3.76)]28.97}{12.01x + y + 16.00z}.$$

Babrauskas [46] suggests that for wood fuels $s = 5.7$. Harmathy [39] notes that a typical wood would have the chemical formula $CH_{1.455}O_{0\ 645} \cdot 0.233H_2O$, which would result in a value of s of 6.0.

3. Several surveys have been published of mass of combustible materials per unit area for different occupancies [28–31]. Given that fire loading can vary significantly over the life of a building, uncertainty should be carefully considered.
4. Properties of liquid fuels: *SFPE Handbook of Fire Protection Engineering* [32].

5. Thermal properties: *SFPE Handbook of Fire Protection Engineering* [27] or manufacturer's data.
6. Building characteristics can be obtained from surveys of existing buildings or architectural plans of new buildings.
7. For design purposes, a value of 0.9 should be assumed for b_p since this would result in the most conservative prediction of T. θ_5 is only relevant if the theoretical heat of combustion is used. If an effective heat of combustion is used, e.g., "chemical" heats of combustion from Tewarson [33], $\theta_5 = 1.0$.

Validation and Limitations

See Appendix B for comparisons of predictions with test data.

Predictions using Babrauskas' method bounded the average temperatures measured in the CIB tests for ventilation-controlled fires but underpredicted average temperatures for fuel-controlled fires. For ventilation-controlled fires, Babrauskas' burning rate prediction falls in the scatter of points. The burning rate data can be bounded by multiplying Babrauskas' prediction by a factor of 1.3 and by dividing by a factor of 2.3.

Babrauskas' method reasonably predicted peak temperatures but underpredicted burning duration in all of the Cardington tests; however, predictions of burning rate improved as $\frac{A}{A_o\sqrt{H_o}}$ increased.

Ma and Mäkeläinen

Ma and Mäkeläinen developed a parametric temperature-time curve for compartments that are small or medium in size (floor area <100 m^2). The method was developed for use mainly with cellulosic fires. Their aims were to develop a simple calculation procedure that would reasonably estimate the temperature, with time, of a fully developed compartment fire.

Ma and Mäkeläinen noted that fires generally only impact the structures during the fully developed and decay stages. They developed a general shape function to define the temperature history of a compartment fire that is a function of fuel loading, ventilation conditions, and geometry and material properties of the compartment.

The general shape function was developed by non-dimensionalizing temperature–time data from 25 different data sets and was based on the maximum gas temperature, T_{gm}, and the time to reach the maximum temperature, t_m. The non-dimensionalized data collapses to the general shape shown in Fig. 9.

The shape of the curve is determined using the following equation and an appropriate value for the shape constant, δ. The recommended values for the shape constant are 0.5 for the ascending phase and 1.0 for the decay phase. These values produce a curve that encompasses a majority of the experimental data. It is reported, however, that values for the shape factor of 0.8 for the ascending phase and 1.6 for

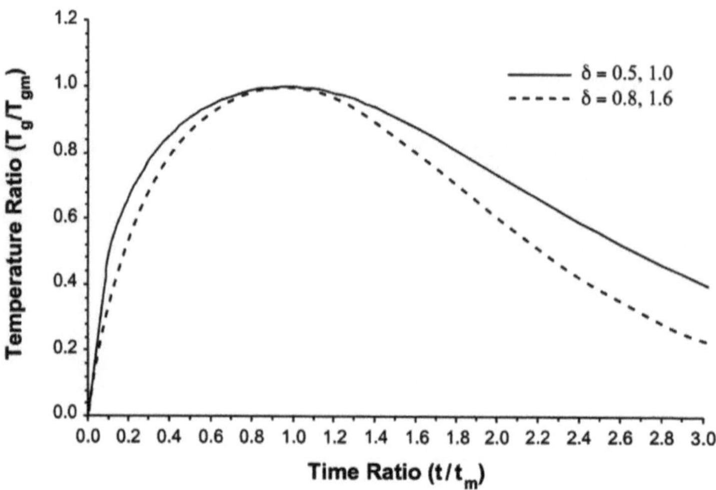

Fig. 9 Non-dimensionalized temperature–time curves developed by Ma and Mäkeläinen [47]

the descending phase provided a best-fit curve to the data [47]. Both curves are shown in Fig. 9.

$$\frac{T - T_o}{T_{gm} - T_o} = \left(\frac{t}{t_m} \exp\left(1 - \frac{t}{t_m}\right)\right)^{\delta}$$

where:

T = Temperature in compartment (°C)
T_o = Ambient temperature (°C)
T_{gm} = Maximum temperature in compartment (°C)
t = Time (min)
t_m = Time corresponding to maximum gas temperature (min)
$$t_m = \frac{0.63 m_f}{\dot{m}_f}$$
m_f = Mass of fuel (kg)
$\dot{m}_f$ = Mass burning rate of fuel (kg/min)
δ = Appropriate shape constant of the temperature–time curve discussed above

For *ventilation-controlled* fires, Ma and Mäkeläinen use Law's correlation to describe the duration of the fully developed stage:

$$\dot{m}_f = 10.8\left(1 - \exp\left(-0.036\frac{A}{A_o\sqrt{H_o}}\right)\right)A_0\sqrt{\frac{H_o W}{D}} \; (\text{kg/min})$$

$$T_{gm} = 1240 - 11\frac{A}{A_o\sqrt{H_o}} \; (°\text{C})$$

where:

A = Surface area of interior of enclosure (m^2)
A_o = Area of ventilation opening (m^2)
H_o = Height of ventilation opening (m)
D = Depth of compartment (m)
W = Width of wall containing ventilation opening (m)

For fuel-controlled fires, Ma and Mäkeläinen use Harmathy's correlation for the burning rate of fuel-controlled fires:

$$\dot{m}_f = 0.0062 \frac{A_f}{m_f} m_f$$

where:

A_f = Surface area of fuel (m^2)

For furniture, the value for Af/mf is generally between 0.1 and 0.4 m^2/kg; however, the most common value is between 0.12 and 0.18 m^2/kg, and 0.131 represents the value obtained from a series of Japanese tests [47]. The maximum gas temperature is determined using

$$T_{gm} = T_{gmcr} \sqrt{\frac{A}{\eta_{cr} A_o \sqrt{H_o}}}$$

with the maximum fire temperature in the critical region, T_{gmcr}, determined by

$$T_{gmcr} = 1240 - 11\eta_{cr}$$

and the value of η_{cr} determined using

$$\eta_{cr} = \frac{\rho_o \sqrt{g}}{0.263 \left(\frac{A_f}{m_f} \right) m_f'' \kappa} = \frac{14.34}{\kappa \left(\frac{A_f}{m_f} \right) m_f''}$$

Where:

A_f = Surface area of fuel (m^2)
g = Gravitational constant (9.81 m/s^2)
$\kappa = \frac{A_{floor}}{A}$ = Ratio of floor area to the total compartment surface area
m_f'' = Mass of fuel per unit area (kg/m^2)
ρ_0 = Density of air (kg/m^3)

The shape function is based on 25 experimental data sets whose key parameters, fuel load density, ventilation factor, thermal boundary properties, and room dimensions varied between experimental studies. The ranges for each of these parameters are listed in Table 2.

Table 2 Range of values for key parameters from the 25 data sets used to develop the shape function

Property	Range	Units
Fuel load density, m_f''	10–40	kg/m^2
Ventilation Factor, $A_o\sqrt{H_o}$	5–16	m$^{5/2}$
$\sqrt{k\rho c}$	555–1800	J/m^2 s$^{1/2}$ K
Compartment floor area, A_{floor}	< 100	m^2
Maximum height, H	< 4.5	m
Shape of compartment (W/D)	0.5–2.0	

Data Requirements

1. Ratio of floor area to total surface area
2. The mass of fuel per unit area, m_f''
3. The area and height of the enclosure opening(s), A_o and H_o
4. The interior surface total area of the enclosure, including the floor and openings, A, and the width, W, and depth, D, of the enclosure
5. The surface area-to-mass ratio of the fuel, A_f/m_f

Data Sources

1. The surface area-to-mass ratio of the fuel typically varies between 0.1 and 0.4 m^2/ kg for larger wood cribs and conventional furniture and more often varies between 0.12 and 0.18 m^2/kg [40].
2. Several surveys have been published of mass of combustible materials per unit area for different occupancies [28–31]. Given that fire loading can vary significantly over the life of a building, uncertainty should be carefully considered.
3. Building characteristics can be obtained from surveys of existing buildings or architectural plans of new buildings.

Validation and Limitations

See Appendix B for comparisons of predictions with test data.

Predictions made using Ma and Mäkeläinken's method's maximum temperature predictions bounded the average temperatures measured in the CIB tests for ventilation-limited fires but underpredicted average temperatures for fuel-limited fires. Given that maximum temperature predictions using Ma and Mäkeläinken's method were compared to the CIB data, which represented the average temperatures measured during the fully developed stage, and predictions of average temperature would be lower than average temperatures, Ma and Mäkeläinken's method would underpredict much of the CIB temperature data.

Ma and Mäkeläinken's method reasonably predicted average temperatures and duration in Cardington tests #1, 2, 8, and 9 ($\frac{A}{A_o\sqrt{H_o}} = 16$ to 18 m$^{1/2}$); however, as

$\frac{A}{A_o\sqrt{H_o}}$ increased, predictions increasingly deviated from the test data. See the conclusions regarding Harmathy's and Law's methods for an evaluation of burning rate predictions.

CIB

In 1958, under the auspices of CIB WO 14, laboratories from several countries agreed to investigate the factors that influence the development of enclosure fires [48]. Compartments with dimension ratios of 211, 121, 221, and 441 (where the first number denotes compartment width, the second number denotes compartment depth, and the last number denotes compartment height) with length scales of 0.5 m, 1.0 m, and 1.5 m were analyzed. A total of 321 experiments were conducted in still-air conditions. The fuel loading $\left(m_f''\right)$ in the compartments ranged from 10 to 40 kg/m^2 of wood cribs with stick spacing to stick width ratios of 1/3, 1, and 3. Test data was modified through statistical analysis to account for systematic differences between test laboratories.

Average temperature and normalized burning rate were presented as a function of $\frac{A_o}{A_o\sqrt{H_o}}$ in graphical form (A was defined to exclude the area of the ventilation opening and the floor area). Separate graphs were presented for cribs with 20 mm thick wood sticks spaced 20 mm apart, and for cribs with 20 mm wide sticks spaced 60 mm apart, or with 10 mm wide sticks spaced 30 mm apart. Because the cribs with 20 mm thick wood sticks spaced 20 mm apart resulted in higher compartment temperatures and lower normalized burning rates (and, hence, longer predicted burning durations), these graphs are recommended for design analysis and are presented here.

Figure 10 shows the average compartment temperature during the fully developed burning stage, where "fully developed burning" was defined as the period where the

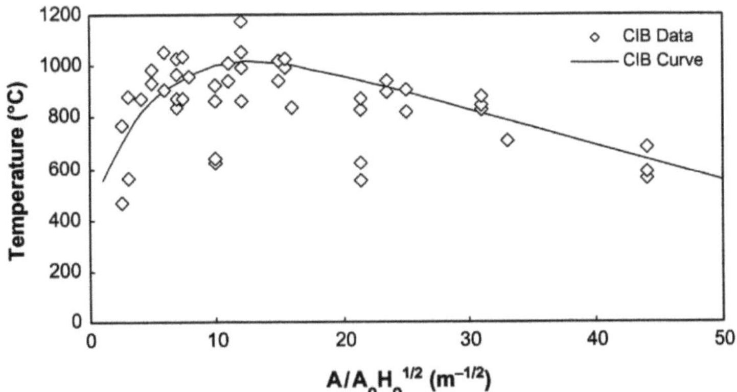

Fig. 10 Average temperature during fully developed burning

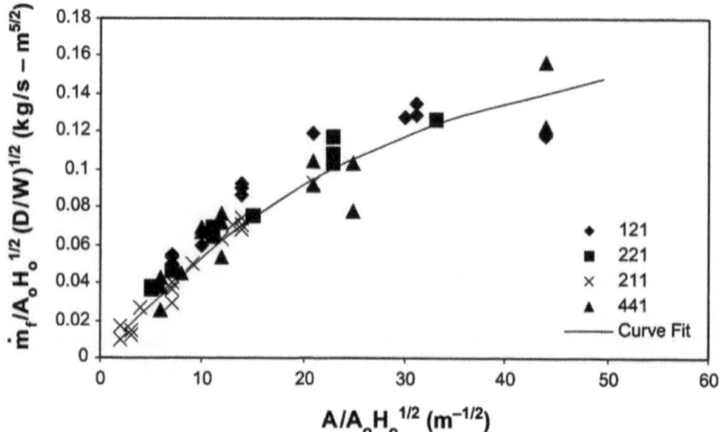

Fig. 11 Normalized burning rate during fully developed burning

mass of fuel was between 80% and 30% of the original, unburned fuel mass. The line represents a best fit through the data.

Figure 11 shows the burning rate, $\dot{m}_f$, during the fully developed burning stage, normalized by the ventilation factor $A_o\sqrt{H_o}$ and the square root of the ratio of compartment depth to width (where the width is the dimension of the wall containing the ventilation opening).

To apply these graphs in a design context, first calculate the factor $\frac{A}{A_o\sqrt{H_o}}$ and use Fig. 10 to determine the average gas temperature. Then, use Fig. 11 to determine the normalized burning rate. The normalized burning rate can be re-dimensionalized by multiplying the ventilation factor $A_o\sqrt{H_o}$ and dividing the square root of the ratio of compartment width to depth. The duration of burning can be determined by dividing the total mass of fuel, m_f, by the burning rate.

Data Requirements

1. The total mass of fuel, m_f
2. The area and height of the enclosure opening(s), A_o and H_o
3. The interior surface total area of the enclosure, excluding the floor and openings, A, and the width, W, and depth, D, of the enclosure

Data Sources

1. Several surveys have been published of mass of combustible materials per unit area for different occupancies [28–31]. Given that fire loading can vary significantly over the life of a building, uncertainty should be carefully considered.
2. Building characteristics can be obtained from surveys of existing buildings or architectural plans of new buildings.

Validation and Limitations

See Appendix B for comparisons of predictions with test data.

The averaged Cardington temperature data falls in the same range as the CIB temperature data for values of $\frac{A}{A_o\sqrt{H_o}}$ less than 30. However, once the opening factor exceeds 30 m$^{-1/2}$, the CIB temperature graph underpredicts temperature, and the CIB data has much lower values than the Cardington data.

As a curve fit through data, the CIB temperature graph reasonably predicted the aggregate of all CIB temperature and burning rate data but underpredicted some experiments and overpredicted others.

Using the CIB graphs resulted in reasonable predictions of average temperature and duration in Cardington tests #1, 2, 8, and 9 and reasonable prediction of duration but underprediction of temperature in Cardington tests #3 and 4. In test #7, using the CIB graphs resulted in reasonable predictions of duration but underprediction of temperatures. Due to the large values of $\frac{A}{A_o\sqrt{H_o}}$ in Cardington tests #5 and 6, predictions were not possible using the CIB graphs.

Law

Law derived a method of predicting compartment temperatures resulting from fully developed fires based on data from tests conducted under the auspices of CIB. Law's method takes into account the geometry of the compartment. The area of the compartment's lining surface through which heat is lost is expressed by subtracting the vent area from the total interior compartment surface area $(A - A_o)$; the temperature in the compartment is therefore dependent on A, as well as variables incorporated in the ventilation factor, A_o, and H.

Law derived the following equation to determine the maximum temperature of the compartment with natural ventilation [49]:

$$T_{gm} = 6000 \left(\frac{1 - e^{-0.1\frac{A}{A_o\sqrt{H_o}}}}{\sqrt{\frac{A}{A_o\sqrt{H_o}}}} \right) (^{\circ}\text{C})$$

where:

T_{gm} = Maximum compartment temperature ($^{\circ}$C)
A = Surface area of interior of enclosure (m^2)
A_o = Area of ventilation opening (m^2)
H_o = Height of ventilation opening (m)

This equation does not account for the effects on compartment temperature due to fuel loading. It simply represents the maximum temperature achieved in a compartment for a given geometry and ventilation. The following equation incorporates the effect of fuel loading on the temperature and is valid for wood-based fuels:

$$T = T_{gm}\left(1 - e^{-0.05\Psi}\right)(°C)$$

where:

$$\Psi = \frac{m_f}{\sqrt{A \times A_o}}$$

m_f = Mass of fuel (kg)

The mass loss rate is correlated as

$$\dot{m}_f = 0.18 A_o \sqrt{\frac{H_o W}{D}} \left(1 - e^{-0.036\frac{A}{A_o\sqrt{H_o}}}\right) \ (kg/s)$$

$$\text{For } \frac{\dot{m}_f}{A_o\sqrt{H_o}} \left(\frac{D}{W}\right)^{\frac{1}{2}} < 60$$

where:

$\dot{m}_f$ = Mass burning rate of fuel (kg/s)
W = Length of wall containing ventilation opening (m)
D = Depth of compartment (m)

The duration of burning can be calculated by dividing the total mass of combustibles by the burning rate as follows:

$$\tau = \frac{m_f}{\dot{m}_f}\,(s)$$

where:

τ = Burning duration (s)

Data Requirements

1. The total mass of fuel, m_f
2. The area and height of the enclosure opening(s), A_o and H_o
3. The interior surface total area of the enclosure, including the floor and openings, A, and the width, W, and depth, D, of the enclosure

Data Sources

1. Several surveys have been published of mass of combustible materials per unit area for different occupancies [28–31]. Given that fire loading can vary significantly over the life of a building, uncertainty should be carefully considered.
2. Building characteristics can be obtained from surveys of existing buildings or architectural plans of new buildings.

Validation and Limitations

See Appendix B for comparisons of predictions with test data.

Without applying the adjustment factor W, Law's temperature predictions bounded all the CIB data. Law reasonably predicted the CIB burning rate data, and, if the burning rate was adjusted by a factor of 1.4, Law bounded all the CIB burning rate data.

Law reasonably predicted average temperature and duration for Cardington tests #1, 2, 8, and 9

$$\left(\frac{A}{A_o \sqrt{H_o}} = 16 \text{ to } 18 \text{ m}^{-1/2} \right).$$

For Cardington tests #3, 4, 5, and 6,

$$\left(\frac{A}{A_o \sqrt{H_o}} = 45 \text{ to } 345 \text{ m}^{-1/2} \right),$$

Law reasonably predicted duration but underpredicted temperature. In test #7

$$\left(\frac{A}{A_o \sqrt{H_o}} = 20 \text{ m}^{-1/2} \right),$$

which was square in plan view, Law's method underpredicted temperatures but predicted the burning duration.

Simple Decay Rates

Many of the methods cited previously do not contain a method of estimating the compartment temperature during the decay stage. For these methods, a number of simple decay rates can be applied if the engineer wishes to account for heating that occurs during the decay phase.

Decay cannot be modeled by basic physics because the "decay rate" is actually the heat transfer from the compartment and the heat release rate of combustibles that

Table 3 Rate of decrease in temperature

Temperature decay (°C/min)	Restrictions	Reference
10	$\tau < 60$ min	Kawagoe
7	$\tau > 60$ min	Kawagoe
>10	$\tau < 60$ min	Magnusson and Thelandersson
<10	$\tau > 60$ min	Magnusson and Thelandersson
10	No restrictions	Swedish Building Regulations
15–20	Short-duration fires	Harmathy

have charred and collapsed onto the floor, with poor access to oxygen and therefore limited heat release rate. All methods are wholly empirical.

The simplest way to determine the temperature time profile during the decay phase is to use a fixed rate of temperature decay. Originally, the temperature decay during the cooling phase was selected arbitrarily. Kawagoe first suggested that the rate of temperature decrease during the cooling period was a function of the fire duration, reporting values of 7 °C/min for fire durations greater than 60 min and 10 °C/min for fire durations less than 60 min [50]. The pioneering work of Magnusson and Thelandersson indicated that the rate of decrease in temperature was a function of the fire duration and opening factor [51–53]. Magnusson and Thelandersson reported that for shorter duration fires the rate of temperature decrease was higher than 10 °C/min, while for longer duration fires the rate of temperature decrease was lower than 10 °C/min [38]. Based on a series of short-duration fires, Harmathy reported decay rates between 15 and 20 °C/min, which are consistent with the results presented by Magnusson and Thelandersson. Typical rates found in the literature are listed in Table 3.

In the absence of better information, it would be appropriate to select a decay rate of 7 °C/min for fires with a predicted duration of 60 min or more and a decay rate of 10 °C/min for fires with a predicted duration of less than 60 min, since these rates would result in the slowest decay rates according to the above.

Recommendations

Based on the comparison of predictions to the data from the CIB and Cardington tests, Law's method is recommended for use in all roughly cubic compartments (compartment width to depth ratio within the range of 0.5–2.0) and in long, narrow compartments where $\frac{A}{A_o\sqrt{H_o}}$ does not exceed ≈ 18 m. To ensure that predictions are sufficiently conservative in design situations, the predicted burning rate should be reduced by a factor of 1.4, and the temperature adjustment should not be reduced by the factor 'F'. See Figs. 12, 13, 14, 15, 16 and 17, which show comparisons made using Law's method to the CIB data and to data for Cardington tests

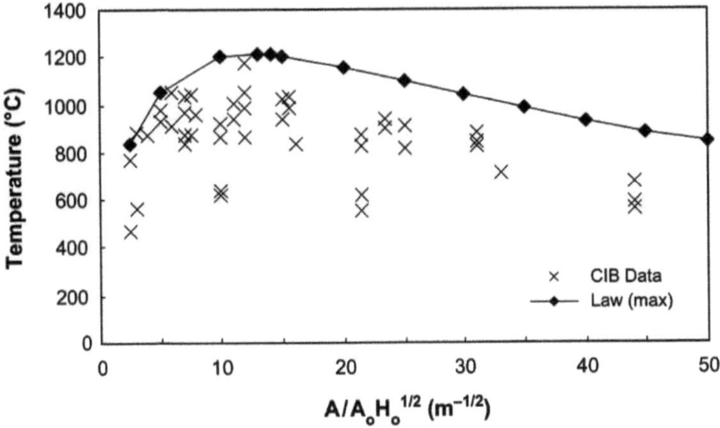

Fig. 12 Comparison of CIB temperature data to predictions using Law's method

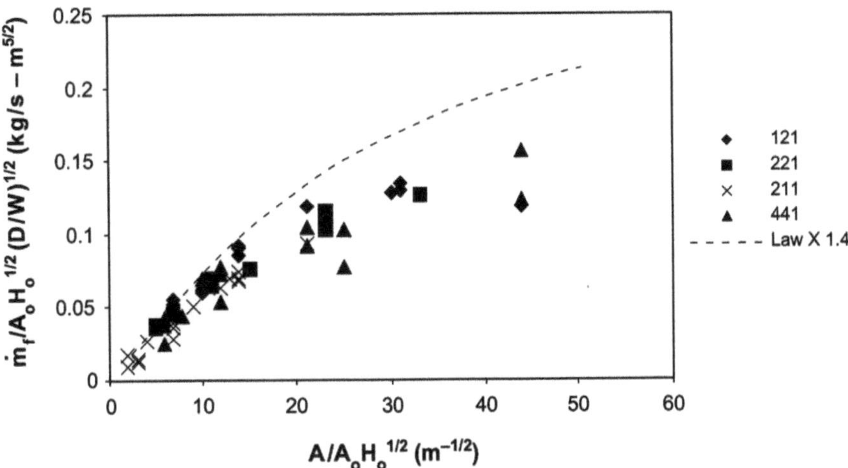

Fig. 13 Comparison of CIB burning rate data to predictions using Law's method

$$\#1 \left(\tfrac{A}{A_o \sqrt{H_o}} = 16 \text{ m}^{-1/2} \right), \quad \#2 \left(\tfrac{A}{A_o \sqrt{H_o}} = 16 \text{ m}^{-1/2} \right), \quad \#8 \left(\tfrac{A}{A_o \sqrt{H_o}} = 18 \text{ m}^{-1/2} \right), \text{ and}$$

$$\#9 \left(\tfrac{A}{A_o \sqrt{H_o}} = 17 \text{ m}^{-1/2} \right).$$

Law's method does not predict temperatures during the decay stage. For cases where a prediction of temperatures during the decay stage is desired, a decay rate of 7 °C/min can be used for fires with a predicted duration of 60 min or more, and a decay rate of 10 °C/min can be used for fires with a predicted duration of less than 60 min.

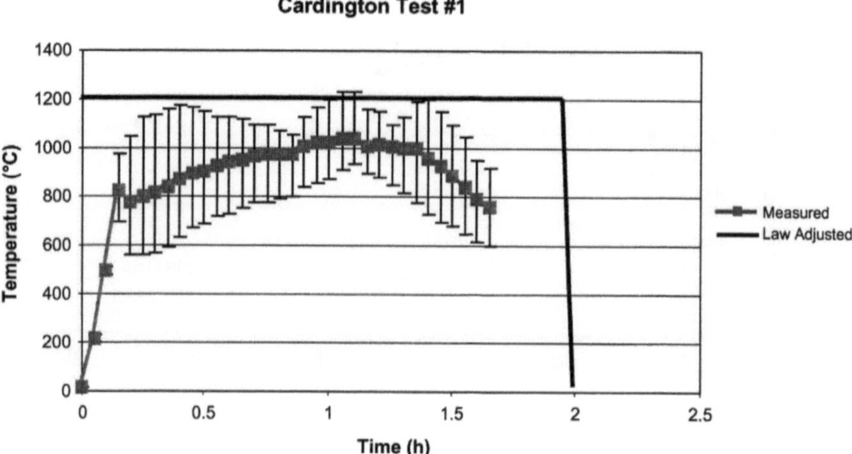

Fig. 14 Comparison of predictions using Law's modified method for Cardington test #1

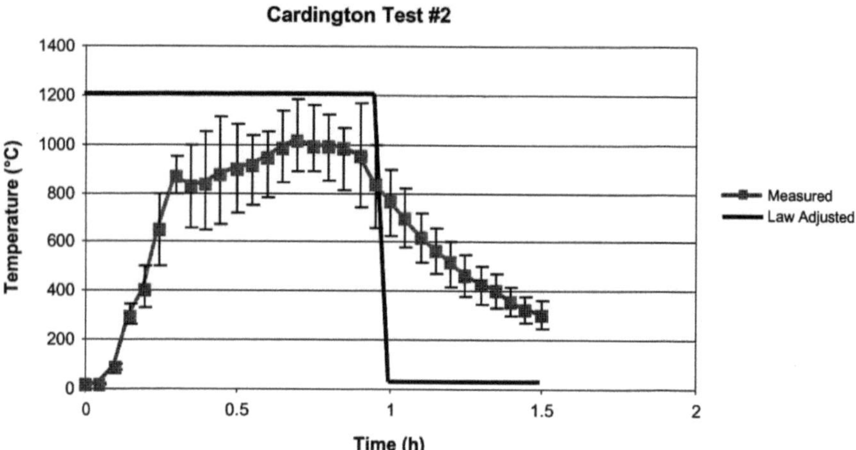

Fig. 15 Comparison of predictions using Law's modified method for Cardington test #2

For long, narrow spaces in which $\frac{A}{A_o\sqrt{H_o}}$ is in the range of 45 to 85 m$^{-1/2}$, Magnusson and Thelandersson provide reasonable predictions of temperature and duration. See Figs. 18, 19 and 20, which show comparisons made using Magnusson and Thelandersson's method to data for Cardington tests #3 $\left(\frac{A}{A_o\sqrt{H_o}} = 45 \text{ m}^{-1/2}\right)$, #4 $\left(\frac{A}{A_o\sqrt{H_o}} = 45 \text{ m}^{-1/2}\right)$, and #5 $\left(\frac{A}{A_o\sqrt{H_o}} = 85 \text{ m}^{-1/2}\right)$.

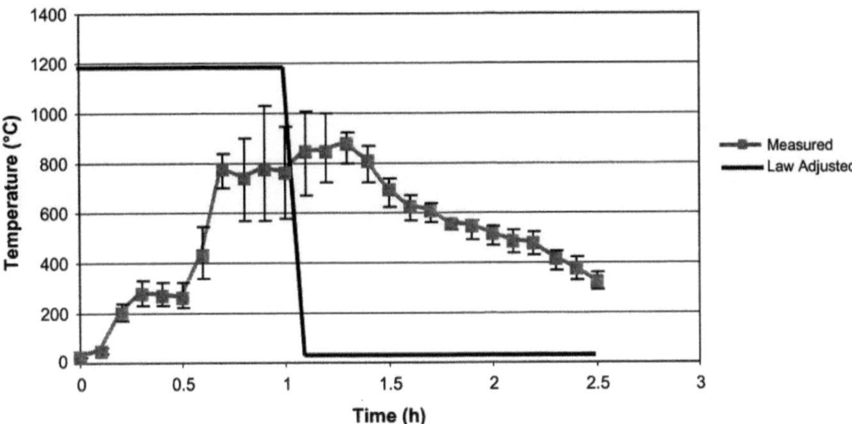

Fig. 16 Comparison of predictions using Law's modified method for Cardington test #8

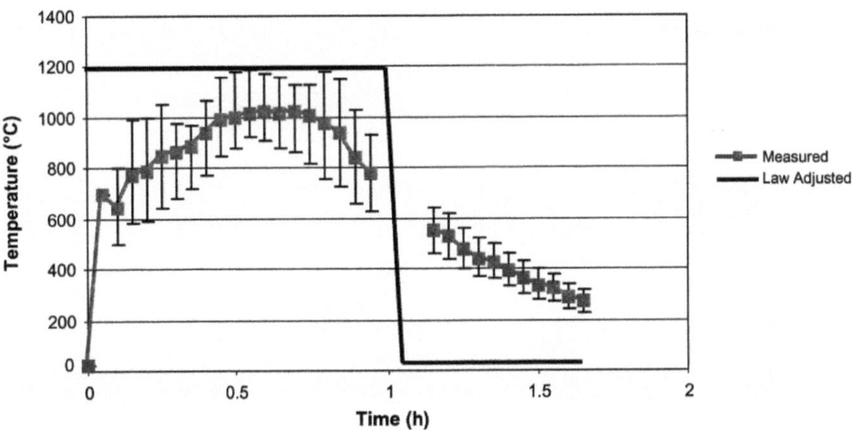

Fig. 17 Comparison of predictions using Law's modified method for Cardington test #9

For long, narrow spaces in which $\frac{A}{A_o\sqrt{H_o}}$ is approximately 345 m$^{-1/2}$, Lie's method is recommended. Note that this value of $\frac{A}{A_o\sqrt{H_o}}$ is outside Lie's stated range of applicability. However, based on the comparison of predictions to the Cardington data, its use is still recommended. See Fig. 21, which shows comparisons made using Lie's method to data for Cardington test #6 $\left(\frac{A}{A_o\sqrt{H_o}} = 345 \text{ m}^{-1/2}\right)$.

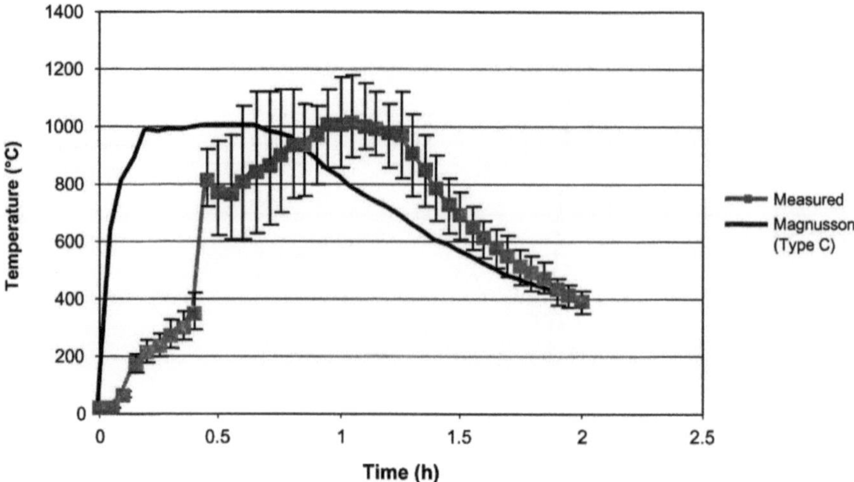

Fig. 18 Comparison of predictions from Magnusson and Thelandersson's method (Type C) to data for Cardington test #3

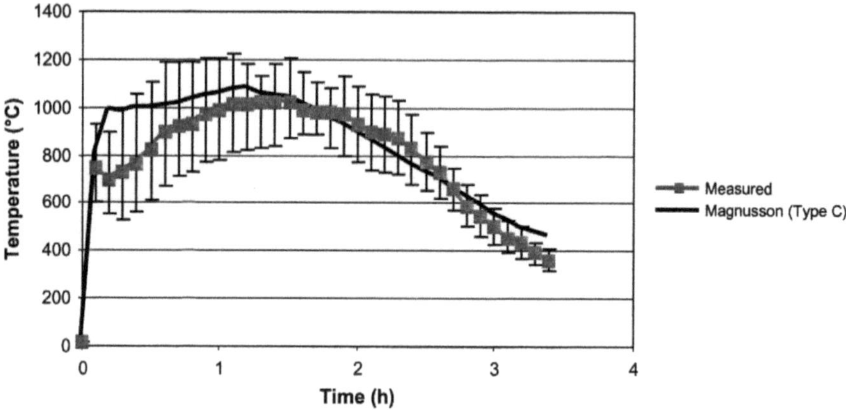

Fig. 19 Comparison of predictions from Magnusson and Thelandersson's method (Type C) to data for Cardington test #4

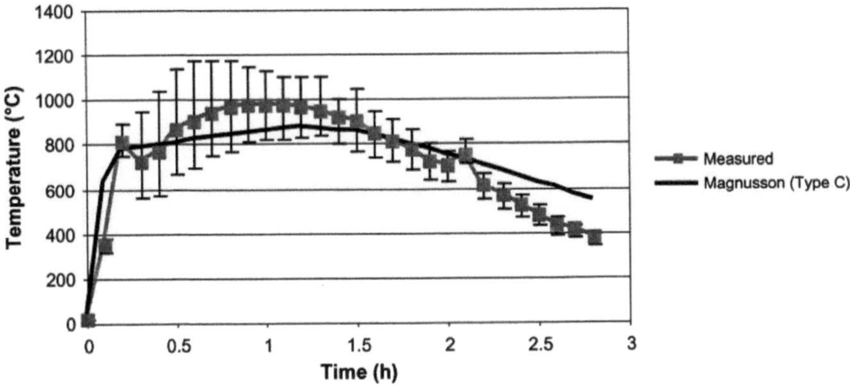

Fig. 20 Comparison of predictions from Magnusson and Thelandersson's method (Type C) to data for Cardington test #5

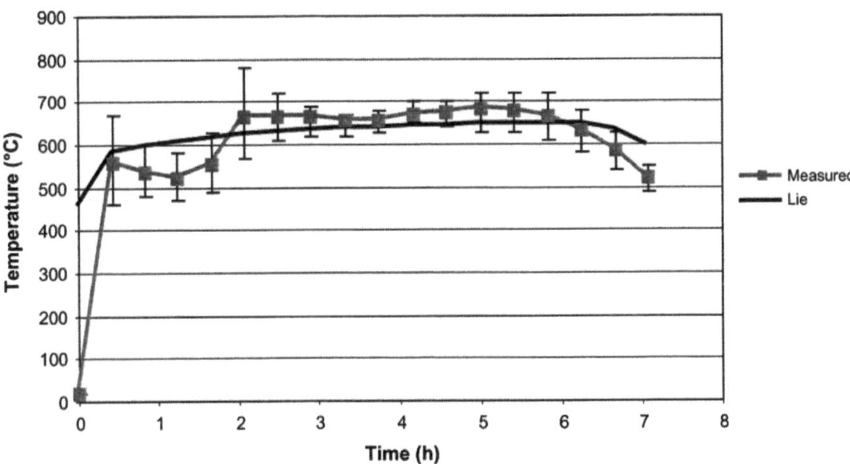

Fig. 21 Comparison of predictions from Lie for Cardington test #6

Fire Exposures from Plumes

This section of the guide focuses on predicting the heat transfer from area exposure fire plumes to adjacent surfaces. Area exposure fires are burning objects or fuel located adjacent to or near the surface being heated. For certain scenarios, the local fire exposure may produce a more extreme exposure than the hot gas layer that develops in the area of consideration. Some examples are open parking garages, large warehouses, and bridges and overpasses. To analyze these scenarios, one needs to have knowledge of the incident heat flux levels produced by local fire plumes.

The boundary condition between the fire plume and the structural element needs to be properly defined in order to predict the temperature of the structural element with time. This part of the guide develops heat transfer boundary conditions for two different types of exposure:

1. Bounding or elements immersed in a fire plume
2. Specific geometries or specific element shapes and orientations

Detailed modeling of the fire from first principles can also be conducted to predict the boundary condition; however, this type of analysis is not addressed in this guide. If detailed modeling is conducted, the model should be verified with existing data for similar configurations to validate predicted heat fluxes. Some additional data on gas temperatures and velocities generated by a fire plume are also included to aid in this type of modeling effort.

The fire exposure recommended for a bounding analysis will consist of constant fire exposure. If a more refined analysis is required, guidance is provided on how to predict the boundary condition with the fire in specific geometries. These geometries include the following:

- Flat vertical walls
- Comers with a ceiling
- Unbounded flat ceilings
- I-beam mounted below a ceiling

Fire Exposures to Structural Elements, The Society of Fire Protection Engineers Series,
https://doi.org/10.1007/978-3-031-64025-4_3

The boundary condition in these configurations is based on experimental data and may be limited to the conditions tested in the study.

Studies have also been conducted to measure the heat flux boundary condition with fires in other configurations. Lattimer [54] provides a review of existing incident heat flux data and correlations for exposure fires and burning surfaces in a variety of configurations including flat walls, comers, corners with a ceiling, parallel flat walls, walls above a window containing a fire plume, unbounded ceiling, and an I-beam under a ceiling.

Axisymmetric Fire Plumes

The simplest fire plume is the unconfined axisymmetric fire plume, shown in Fig. 1. Correlations for velocity and temperature produced by an axisymmetric plume are provided in this section to aid those in modeling the heat flux to elements from first principles. Unconfined axisymmetric fire plumes are typically approximated as point heat sources when estimating the local velocity and temperature profile. This section describes how to estimate the location of the virtual point source relative to the base of the fire, the flame height, and the velocity and temperature distribution within the fire plume.

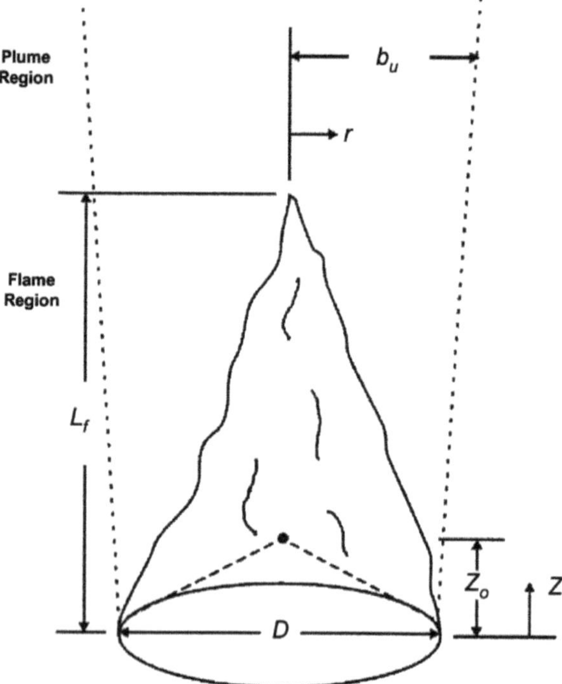

Fig. 1 Axisymmetric fire plume

Velocity Profile

The velocity profile of the fire plume is a function of the elevation above the virtual origin and the distance from the die plume centerline. The velocity at the plume centerline can be calculated using the following equation for regions above the average flame height, L_f: [55]

$$z > L_f, \; U_{m,c}(Z) = 1.14 \left[(1 - \chi_r)\dot{Q} \right]^{1/3} (Z - z_o)^{-1/3} \; (\mathrm{m/s}) \tag{1}$$

where:

$U_{m,c}(Z)$ = Centerline plume velocity (m/s)
χr = Fraction of energy released as radiation in the fire
$\dot{Q}$ = Fire heat release rate (kW)
Z = Target elevation above the base of the fire (m)
z_o = Elevation of the virtual origin relative to the base of the fire (m)

The centerline plume velocity for regions below the average flame height may be determined using Eq. 2 [56] where all terms have been defined.

$$Z < L_f, \; U_{m,c}(Z) = 6.83 Z^{0.5} \left(0.02 < Z/\dot{Q}^{2/5} < 0.08 \right) \; (\mathrm{m/s}) \tag{2}$$

The virtual origin may be calculated using Eq. 3 where D is the effective fire diameter (m) [57]:

$$z_o = -1.02D + 0.083\dot{Q}^{0.4} \; (\mathrm{m}) \tag{3}$$

For noncircular fuel packages with a length-to-width ratio of near one, the equivalent diameter of the fuel package can be estimated using the surface area, A, of the noncircular fuel package:

$$D = \sqrt{\frac{4A}{\pi}} \; (\mathrm{m}) \tag{4}$$

where:

A = Surface area of the fuel package (m^2)

The average flame height can be calculated using the relation developed by Heslcestad [58]:

$$L_f = 0.23\dot{Q}^{2/3} - 1.02D \; (\mathrm{m}) \tag{5}$$

where:

$\dot{Q}$ = *Heat release rate of the fire (kW)*
D = *Diameter of the fuel package (m)*

The velocity distribution within a fire plume has been found to fit a Gaussian profile, though no theoretical grounds exist for this [55].' [59] The following equation may be used to determine the velocity as a function of the distance from the plume centerline [55]:

$$U(r) = U_{m,c}(Z) \cdot \exp\left(\frac{-r^2}{b_u^2}\right) \quad \text{(m/s)} \tag{6}$$

where:

$U(r) =$ Velocity in plume at a distance r (m) from the centerline (m/s)
$b_u =$ Plume width parameter (m)

The plume width parameter is found via Eq. 7 where all terms have been defined [55].

$$b_u = 0.13(Z - z_o) \quad \text{(m)} \tag{7}$$

Temperature Profile

The temperature profile is also a function of the elevation above the plume's virtual origin and the distance from the plume centerline. The centerline temperature may be calculated using Eq. 8 for elevations above the average flame height [55]:

$$Z > L_f, T_{m,c}(Z) = 26[(1 - \chi_r)Q]^{2/3}(Z - z_o)^{-5/3} + T_\infty \quad \text{(K)} \tag{8}$$

where:

$T_{m,c}(Z) =$ Centerline plume temperature (K)
$T =$ Ambient temperature (K)

The centerline plume temperature for elevations below the average flame height may be determined using the following where all terms have been defined [56]:

$$Z < L_f, T_{m,c}(Z) = 980 + T_\infty \quad \left(\dot{Q}^{2/5} > 10.2D\right) \text{ and } \left(0.03 < Z/\dot{Q}^{2/3} < 0.08\right) \quad \text{(K)} \tag{9}$$

$$T_{m,c}(Z) = 960 + T_\infty \quad \left(\dot{Q}^{2/5} > 10.2D\right) \text{ and } \left(0.02 < Z/\dot{Q}^{2/3} < 0.06\right) \quad \text{(K)} \tag{10}$$

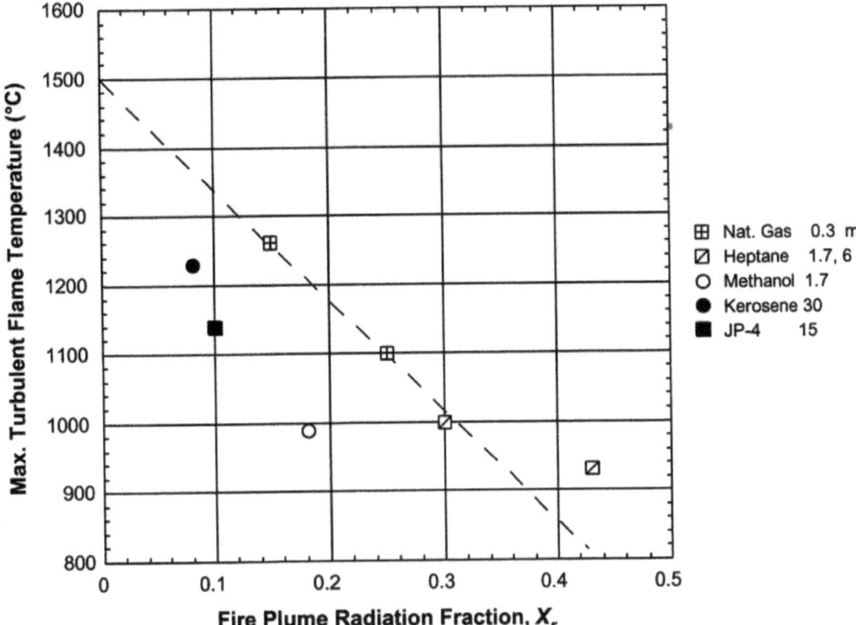

Fig. 2 Maximum turbulent fire plume temperatures from various sources [61–64]

These temperatures represent average temperatures in the flaming and plume regions, and they will tend to be higher when the radiative fraction, χ_r, of the fire is decreased. For turbulent fire plumes, having a radiative loss fraction χ_r, the turbulent flame (centerline) temperature follows the relationship [60]

$$T_{f,\mathrm{plume}} - T_o = k_T(1 - \chi_r)\Delta h_{\mathrm{air}}/c_p \tag{11}$$

From the best available data [61–63], the turbulent mixing parameter, k_T, is found to be about 0.5 for $c_p = 1$ kJ/kg-K. As the fire diameter increases, the radiative fraction falls due to soot blockage [64]. Figure 2 shows flame temperature data for turbulent plumes as a function of χ_r. The extrapolated adiabatic temperature is about 1500 °C.

Temperatures have been measured to be as low as 820 °C for flames produced by fuels with a radiative fraction of $\chi_r \sim 0.20$ [65]. Thus, Eqs. 9 and 10 correspond to fires of $\chi_r \approx 0.3$.

As the pool fire diameter is increased, flames produce more soot, reducing the flame radiation being emitted to the surroundings. From the SFPE Engineering Guide on *Assessing Flame Radiation to External Targets from Pool Fires* [66] and Beyler [67], radiative fraction will decrease linearly from an average radiative fraction of 0.22 for a small-(~0 m) diameter pool fire to approximately 0.04 for a 50 m diameter pool fire. Baum and McCaffrey [68] clearly showed the dependence of gas temperature on diameter, with measured gas temperatures as high as 1000 °C for 6 m diameter fires and 1250 °C for 30 m diameter fires. These data are represented in Fig. 2. The temperature distribution as a function of the distance from the plume centerline also fits a Gaussian profile [55]. Equation 12 can be used to determine the temperature at any distance r (m) from the plume centerline [55]:

$$T(r) = (T_{m,c}(Z) - T_\infty) \cdot \exp\left(0.9\frac{r^2}{b_t^2}\right) + T_\infty \quad \text{(K)} \tag{12}$$

where:

$b_t =$ Thermal plume width parameter (m)

The thermal plume width parameter may be calculated using Eq. 13 where all terms have been defined [55]:

$$b_t = 0.176(Z - z_o) \quad \text{(m)} \tag{13}$$

Data Requirements

1. Source fire heat release rate, $\dot{Q}$ (kW)
2. Radiative fraction, χ_r
3. Elevation above source fire, Z (m)
4. Radial (horizontal) separation from the centerline of the source fire, r (m)

Data Sources

1. Heat release rate data may be obtained from Babrauskas [69], Hoglander and Sundstrum [70], or Mudan and Croce [71].
2. Radiant fraction data may be obtained from Tewarson [72].

Assumptions

1. The axisymmetric fire plume may be approximated as a point heat source. This assumption is valid for many types of fires including pool fires but may yield poor results for three-dimensional burning objects (i.e., sofa), momentum-driven plumes (jets), or regions near the base of the fire.

2. The effect of a hot smoke layer formation in a compartment on the temperature and velocity profiles in a fire plume is ignored. Refer to Evans [73] and Cooper [74] for a discussion of hot layer-plume interactions.
3. There is no air movement (wind, vent flows) in the vicinity of the plume. Such air motions may cause a plume to deflect.

Validation

There have been numerous experiments on the centerline temperature and velocity in fire plumes. The form of the correlations is generally identical; however, there is some variation among the correlated constants [55, 75]. Those presented in this section tend to be conservative in terms of predicting the greatest velocity and centerline temperature for a given heat release rate and target elevation.

Limitations

The fire plume equations in this section are limited to open, axisymmetric thermal plumes in a quiescent environment. The source fire should have a relatively square plan area, though fuel packages or source fires with aspect ratios on the order of two or three may be acceptable. Larger aspect ratios could result in a line fire. Refer to Quintiere and Grove [76] for a discussion of line fire thermal plumes.

Heat Flux Boundary Condition

The governing boundary condition for a fire heating an adjacent surface is determined using the heat balance and is shown in Fig. 3 to be

$$-k\frac{dT}{dx} = q''_{\text{not}} = \alpha_s \varepsilon_f \sigma T_f^4 + h(T_f - T_s) - \varepsilon_s \sigma T_s^4 \qquad (14)$$

assuming negligible heating from the surrounding environment (i.e., no hot gas layer heating). To apply this relation directly, the local gas temperature, T_f; local heat transfer coefficient, h; and the emissivity of the gases, ε_f, must be known. The surface absorptivity, α_s, and emissivity, ε_s, must also be known, but approach 1.0 as they become soot covered. All these parameters are scenario dependent, and all are not readily known or predicted. As a result, several research efforts have been conducted to measure the total incident heat flux to a surface in a variety of configurations. This is typically done using cooled total heat flux gauges. These gauges are cooled so that their surface temperature remains near ambient and are

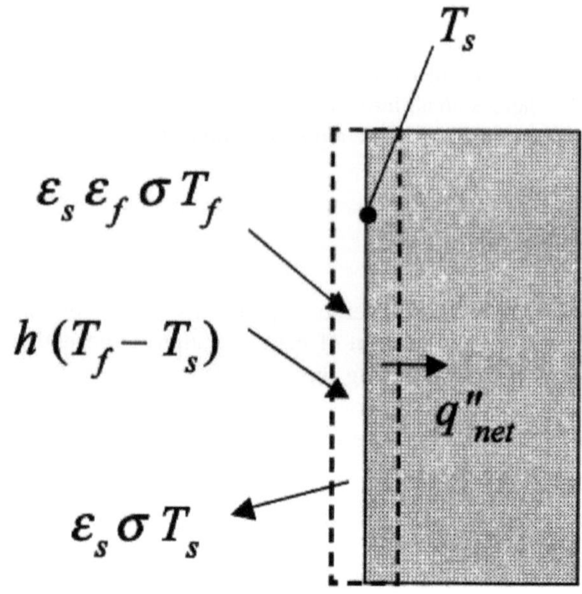

Fig. 3 Heat balance at the material surface

coated with a high-emissivity paint to maximize the absorbed radiation. By setting the surface temperature to the ambient in Eq. 14, the boundary condition at the total heat flux gauge is represented by Eq. 15:

$$q''_{hfg} = \alpha_{s,hfg}\varepsilon_f \sigma T_f^4 + h(T_f - T_\infty) - \varepsilon_{s,hfg}\sigma T_\infty^4 \tag{15}$$

Cooling the gauge surface maximizes the convective heat transfer and minimizes the radiative losses; thus, the cooled heat flux gauges measure the maximum total incident heat flux. Assuming that the surface absorptivity and emissivity are identical, and the emissivity of the heat flux gauge is similar to that of the material surface ($\varepsilon_{s,hfg} \approx \varepsilon_s$), the total incident heat flux measured using the heat flux gauge, Eq. 15, is related to the actual heat flux through the following relation:

$$q''_{net} = \left[\varepsilon_s\varepsilon_f\sigma T_f^4 + h(T_f - T_\infty) - \varepsilon_s\sigma T_\infty^4\right] - h(T_s - T_\infty) - \varepsilon_s\sigma(T_s^4 - T_\infty^4) \tag{16}$$

or

$$q''_{net} = q''_{hfg} - h(T_s - T_\infty) - \varepsilon_s\sigma(T_s^4 - T_\infty^4) \tag{17}$$

Therefore, measuring the heat flux has removed the need to predict both the gas temperature and the emissivity of the gases. To get the actual net heat flux into the surface from the measured heat flux, a surface temperature correction needs to be applied as done in Eq. 17.

A conservative estimate of the net heat flux into the structural element can be determined by either not applying any surface temperature correction or only applying the radiative correction. A closer estimate of the actual net heat flux into the surface would include both radiative and convective corrections. Applying a convective correction involves estimating the local heat transfer coefficient, A, which is dependent on the local velocity and gas temperature.

Local heat transfer coefficients may range from 0.015 to 0.030 kW/(m K) for hot gas flow up a wall or along a ceiling. At points where hot gases impinge on a surface, this value may be higher. Based on data from Kokkala [77]' [78] and You and Faeth [79, 80], the local convective heat transfer coefficient where a diffusion flame impinges on a ceiling is on the order of 0.050 kW/(m-K). Figure 4 contains plots of the radiative correction for different element surface temperatures along with convective correction for convective heat transfer coefficients of 0.015 and 0.050 kW/(m K). From Eq. 17 and Fig. 4, overestimating the convective correction will result in a non-conservative boundary condition. Therefore, a convective heat transfer correction is only recommended in simple configurations where local heat transfer coefficients can be calculated (e.g., flat walls).

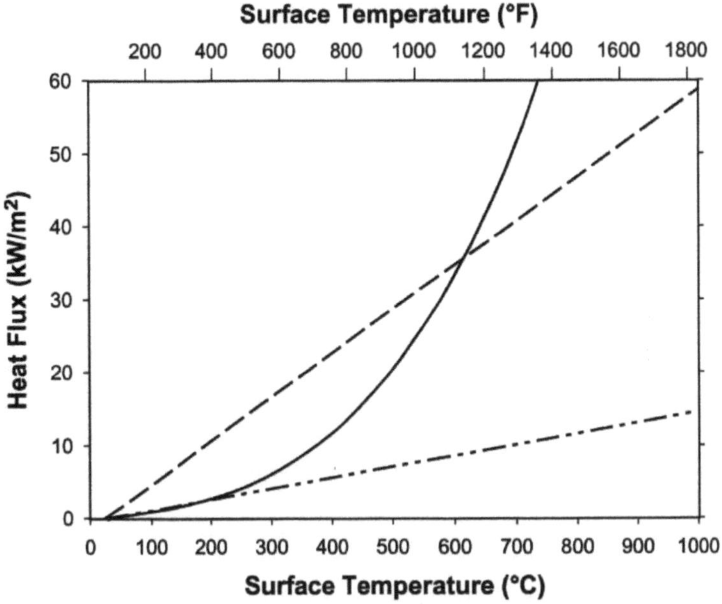

Fig. 4 Magnitude of the surface temperature corrections on the measured total heat flux using a cooled gauge (see Eq. 17). Radiation (—), convection with h = 0.015 kW/(m K) (– ·· –), and convection with h = 0.050 kW/(m K) (– – –)

Bounding Heat Flux: Objects Immersed in Flames

The simplest and most conservative way to treat the fire exposure boundary condition would be to apply a constant, bounding heat flux to all structural elements in the area of interest. The bounding heat flux boundary condition was developed from data on objects immersed in large hydrocarbon pool fires. The heat flux data for objects immersed in fires are presented in this section and used to determine the magnitude of the bounding heat flux. The information in this section is derived primarily from direct or indirect measurements of heat flux taken in open hydrocarbon pool fires with optically thick flames. There is insufficient data available at this time to adequately address the impact of a boundary such as a wall or ceiling on the heat flux conditions of an immersed object. It is expected that the data obtained from optically thick flames in unconfined pool fires is bounding.

Test Data

A series of 30-min, 9.1 m by 18.3 m hydrocarbon pool fires (JP-4) conducted by Gregory, Mata, and Keltner [81] provided useful temperature and heat flux data at various elevations above the base of the fire. Steel cylinders filled or lined with insulation (referred to as small or large calorimeters, respectively) at several locations were used to indirectly measure the net heat flux for objects immersed in the fire. The temperature inside the cylinder was recorded, and the net heat flux was extracted using the inside temperature as a boundary condition. The measurements were taken at various elevations and angular positions on the calorimeters. The cold-wall (i.e., peak) heat fluxes to the large calorimeter varied between 100 kW/m^2 and 160 kW/m^2 at any one location, with the largest peak heat fluxes observed on the underside and the lowest on the top. Figure 5 shows the average peak heat flux at various angular positions as a function of the external surface temperature of the large calorimeter, which increases as a function of time, and the angular position.

The cold-wall fluxes to the small calorimeter varied between 150 kW/m^2 and 220 kW/m^2. As with the large calorimeter data, the maximum heat fluxes were observed on the bottom of the calorimeter, and the minimum were observed on the top. There was no decrease in the cold-wall heat flux detected over the elevation range (1–11 m) sampled.

Russell and Canfield [82] immersed a steel cylinder in a 2.4 m by 4.9 m JP-5 pool fire in windy conditions. The inside surface temperature of the cylinder was directly measured, and the exposure heat flux was determined in the same manner as Gregory, Mata, and Keltner [81]. The peak heat fluxes to the surface of the cylinder were measured at various angular positions. The peak heat fluxes ranged from 18 kW/m^2 on the windward side to 144 kW/m^2 on the leeward side. The heat fluxes on the top and bottom of the cylinder were 48 kW/m^2 and 103 kW/m^2, respectively.

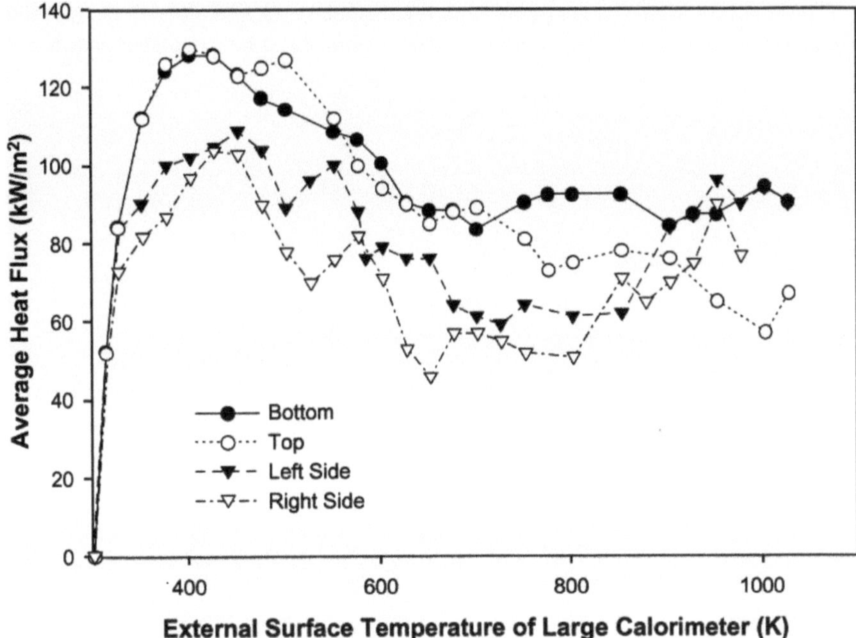

Fig. 5 Averaged peak heat flux as a function of angular position

Table 1 Selected heat fluxes to objects immersed in large pool fires [83]

Test	Pool size	Fuel	Peak heat Flux (kW/m^2)
AEA Winfrith84	0.49 × 9.4 m	Kerosene	150
US DOT84	Not listed	Kerosene	138
USCG84	Not listed	Kerosene	110–142
US DOT84	Not listed	Kerosene	136–159
Sandia84	Not listed	Kerosene	113–150
HSE Buxton84	Not listed	Kerosene	130
Shell Research84	4.0 × 7.0 m	Kerosene	94–112
Large cylinder82	9 × 18 m	JP-4	100–150
Large cylinder82	9 × 18 m	JP-4	150–220
Russell and Canfield83	2.4 × 4.9 m	JP-5	144

Cowley [83] summarized the peak heat fluxes measured directly or indirectly to objects immersed in various large-scale pool fires. The values range between 80 kW/m^2 and 270 kW/m^2. Table 1 summarizes some of this information. Cowley speculates that differences between low- and high-volatile fuels with heat fluxes as high as 300 kw/m^2 are possible in the latter.

Most of the heat flux test data suggest a bounding cold-wall heat flux between 150 kW/m^2 and 170 kW/m^2. Although some data (small calorimeter) indicate that the peak may be as high as 220 kW/m^2, these appear to be exceptional.

The heat flux in a flame increases with fire diameter and where the object or flame impingement is located. The upper bound of heat flux can be calculated as follows:

$$\dot{q}'' = \sigma T_f^4 \qquad (18a)$$

Data Requirements

The flame temperature is needed to perform this calculation.

Data Sources

For pool fires, the radiative fraction can be determined as a function of pool diameter from the SFPE Engineering Guide to *Assessing Flame Radiation to External Targets from Pool Fires*. This radiative fraction can be substituted into Fig. 2 to estimate the flame temperature. For noncircular pools with a length-to-width ratio of near one, the equivalent diameter of the pool can be estimated using the surface area, A, of the noncircular pool:

$$D = \sqrt{\frac{4A}{\pi}} \quad (\mathrm{m}) \qquad (18b)$$

where:

A = Surface area of the fuel package (m^2)

Assumptions

1. The flame emissivity and surface absorptivity are equal to 1.0.
2. The impact of a compartment on the heat fluxes at the surface of an immersed object can be ignored.
3. Reduction in net heat flux due to heating of the target is not considered.

Validation

Equation 18a is based on first principles. Heat fluxes calculated using Eq. 18a are much larger than measured heat fluxes. For example, Baum and McCaffrey [68] reported gas temperatures as high as 1250 °C in 30 m diameter pool fires. Assuming that the gases are optically thick, emissivity of 1.0, the cold-wall heat flux is 305 kW/m^2. As seen in Table 1, measured values are less than this value, indicating that the assumed emissivity may be significantly less than 1.0 or the effective gas temperatures providing the radiation are lower than measured or reported temperatures.

Limitations

The results of this section are limited to Class A (plastic or wood-based) combustible material fires or hydrocarbon pool fires. Gaseous jet flames are beyond the scope of this section because they may produce larger cold-wall (200–270 kW/m^2) heat fluxes to immersed objects [83].

The results are also not applicable to objects that are located near (collocated), but not in, the burning region. Methods of estimating the incident heat flux to collocated objects are available in another Engineering Guide [66].

Heat Fluxes for Specific Geometries

The incident heat flux from a fire plume to a surface is dependent on:

- Geometry
- Dimensions of the fire
- Fire heat release rate
- Effective radiative path length
- Soot production rate

Research has been conducted to evaluate the effects of each of these variables on the incident heat flux from a fire. However, a general engineering approach has not been developed for predicting the incident heat flux from a fire to an adjacent surface. This section provides empirical correlations for estimating the heat flux boundary condition in some specific geometries. These correlations were developed over a specific range of fire source size, heat release rate, and geometry, which limits their general applicability.

The heat transfer from a flame to an adjacent surface or object has historically been characterized with respect to the flame length. Many of the heat flux correlations developed in the literature are based on flame length data taken in a particular study. Measured flame lengths can vary depending on the measurement technique, definition, and surrounding geometry. For the studies considered in this section, the data were nondimensionalized with either the average (50% intermittent) flame length or the flame tip length. Therefore, heat flux correlations should be applied using either the flame length correlation developed in the study or with one that has been demonstrated to predict the flame length in that study.

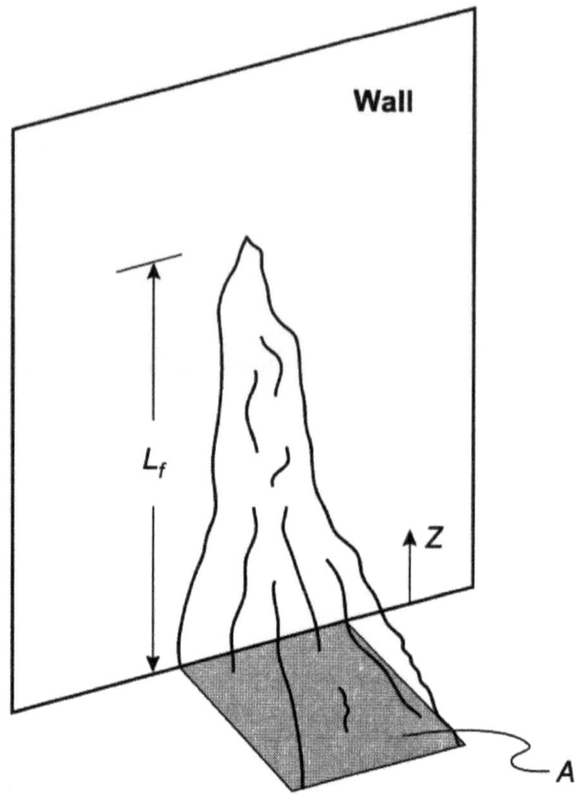

Fig. 6 Fire against a flat vertical wall

Flat Vertical Walls

The simplest geometry is with the fire directly against a flat wall as shown in Fig. 6. Correlations are developed in this section to estimate the vertical and horizontal variation in the heat flux to the wall due to a fire in this configuration.

Correlations to estimate the incident heat flux from an exposure fire against a flat wall have been developed through an experimental study performed by Back et al. [84] In this study, fires were generated using square propane sand burners with edge lengths of 0.28, 0.37, 0.48, 0.57, and 0.70 m. Heat flux fields were measured for fires ranging from 50 to 520 kW. The flame height to burner diameter aspect ratio ranged from approximately 1 to 3 in these tests.

The average flame length of fires against a flat wall was determined to be equal to the average flame length of unconfined fire plumes. Flame lengths can be calculated using the relation developed by Heslcestad [58]:

$$L_f = 0.23\dot{Q}^{2/5} - 1.02D \ (\text{m}) \tag{19}$$

where:

$\dot{Q} =$ Heat release rate of the fire (kW)
$D =$ Diameter of the fuel package (m)

Flame lengths are taken relative to the base of the fire. For noncircular fuel packages with a length-to-width ratio of near one, the equivalent diameter of the fuel package can be estimated using the surface area, A, of the noncircular fuel package:

$$D = \sqrt{\frac{4A}{\pi}} \quad (\mathrm{m}) \tag{20}$$

where:

$A =$ Surface area of the fuel package (m^2)

A plot of the peak heat fluxes measured for each of the fires considered in the study is shown in Fig. 7. Peak heat fluxes for the different fires evaluated were determined to be a function of the fire heat release rate. This dependence was attributed to the larger size fires resulting in thicker boundary layers on the wall, thus increasing the radiation path length. Based on gray-gas radiation theory, the authors found the following relation adequately represented the data:

$$\dot{q}''_{\mathrm{peak}} = 200\left[1 - \exp\left(-0.09\dot{Q}^{1/3}\right)\right] \quad (\mathrm{kW/m^2}) \tag{21}$$

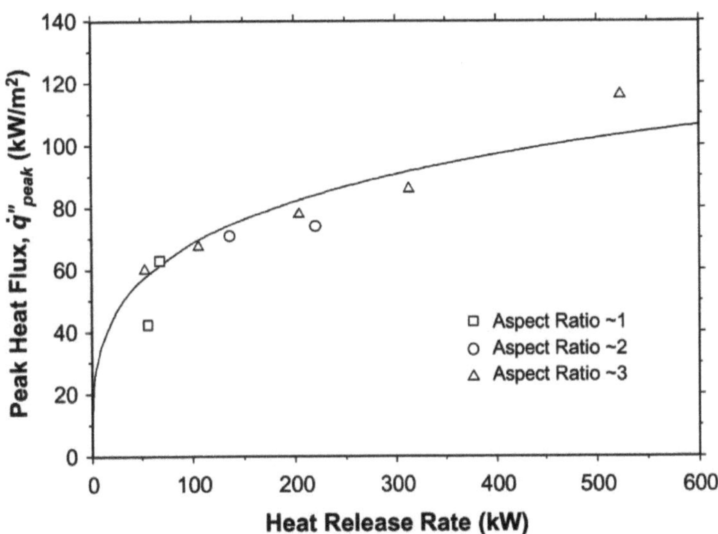

Fig. 7 Peak heat release rates measured in square propane burner fires against a flat wall [84]

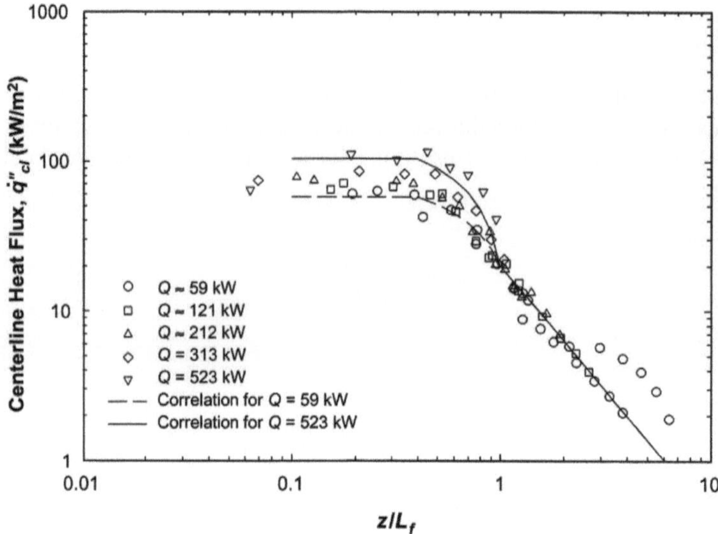

Fig. 8 Vertical heat flux distribution along the centerline of a square propane burner fire adjacent to a flat wall [84]

These peak heat fluxes were measured in the lower part of the fire $(z/L_f \leq 0.4)$ along the centerline. Above this region, the heat fluxes were measured to decrease with distance above the fire, z. The heat flux data measured along the centerline are shown in Fig. 8. Lines in this plot are a general correlation of the centerline data:

$$\dot{q}''_{cl} = \dot{q}''_{peak} \quad z/L_f \leq 0.4 \quad (kW/m^2) \tag{22a}$$

$$\dot{q}''_{cl} = \dot{q}''_{peak} - \frac{5}{3}\left(z/L_f - 2/5\right)\left(\dot{q}''_{peak} - 20\right) \quad 0.4 < z/L_f \leq 1.0 \quad (kW/m^2) \tag{22b}$$

$$\dot{q}''_{cl} = 20\left(z/L_f\right)^{-5/3} \quad z/L_f > 1.0 \quad (kW/m^2) \tag{22c}$$

Heat fluxes were measured by Back et al. [84] to decrease with horizontal distance from the centerline as shown in Fig. 9. Significant heat fluxes were measured as far as twice the burner radius from the centerline. Conservatively, it can be assumed that the heat flux is equal to the centerline heat flux at distances as far as twice the fire radius from the centerline.

Data Requirements

1. Diameter of the fuel package, D. For noncircular fuel packages, the equivalent diameter may be calculated using Eq. 20 and the surface area of the fuel package.
2. Heat release rate of the fire, $\dot{Q}$.
3. Elevation along the flame length, z.

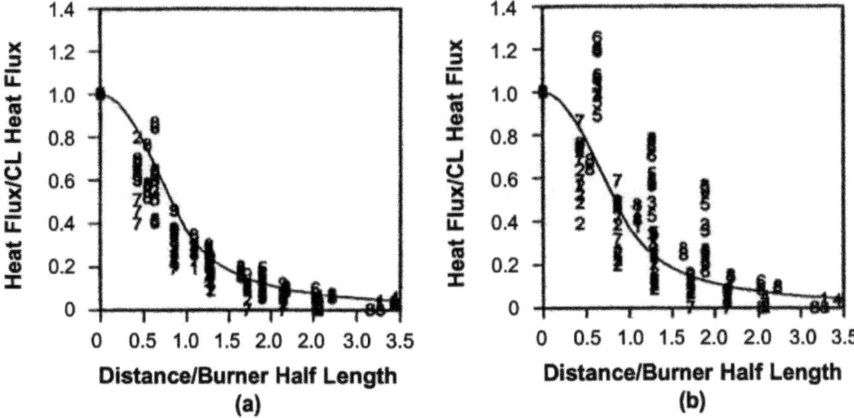

Fig. 9 Horizontal heat flux distribution (**a**) below the flame height and (**b**) above the flame height with distance from the centerline of the fire [84]

Data Sources

1. Heat release rate data may be obtained from Babrauskas [69], Hoglander and Sundstrum [70], or Mudan and Croce [71].

Assumptions

This analysis assumes that the fire is attached to the wall and that the wall is vertical. Walls that are not vertical may result in different total incident heat flux levels due to the flame's becoming separated from the wall or the difference in entrainment into the plume.

Validation

Some studies have made measurements of incident heat fluxes from various burning objects to walls, but the data is sparse. Incident heat fluxes at the rim of wastebasket fires were reported by Gross and Fang [85]. At the rim, heat fluxes as high as 50 kW/m^2 were measured; however, the authors noted that peak heat fluxes for these fires occurred approximately 0.22 m above the rim. Mizuno and Kawagoe [86] performed experiments with upholstered chair fires against a flat wall. In these tests, Mizuno and Kawagoe measured heat fluxes to the wall of 40–100 kW/m^2 over the continuous flaming region ($\sim z/L_f < 0.4$). All these tests were performed using foam-padded chairs. These data do provide evidence that the magnitude of the incident heat flux levels measured in the propane burner experiments is consistent with fire produced by burning items. In tests with propane gas burners against a non-combustible boundary, similar heat flux levels have been measured by other investigators for limited conditions [87, 88].

Limitations

Correlations for incident heat fluxes were developed using luminous flames in an open environment with the fire directly against a flat vertical wall. Using these relations inherently assumes:

- There is negligible heating from a hot gas layer in the surroundings.
- The fire is against the wall.
- The flames are luminous.
- The wall is vertical.

The experimental study considered fire diameters as large as 0.70 m and heat release rates as large as 520 kW. No data was available to validate the correlations against fires with larger diameters or higher heat release rates. The presence of a hot gas layer may increase the total incident flux onto the wall, and if significant in the area of interest, adding this contribution to the total incident heat flux from the fire plume may be warranted [89]. Moving the fire away from the wall will eventually cause the incident heat fluxes to become lower, largely because the flame becomes detached from the wall [90]. Thus, the use of correlations in this section for fires that may be slightly spaced from the wall will yield conservative results. Flames less luminous than those produced by the propane fires (i.e., natural gas) may transmit lower total incident heat fluxes to the wall because the radiative heat flux to the wall will be lower [87, 88]. Propane fuel fires used to develop the heat flux data presented in this section produce a moderate amount of soot; therefore, heat flux levels presented in this section should be considered to be average but not bounding for all different fuels. Propane burners are also used extensively in standard fire tests as an exposure fire that is representative of real fires. Therefore, the incident heat fluxes from these flames are considered to be representative of those produced by most fires.

Fires in a Corner

Fires located in a corner geometry as shown in Fig. 10 produce a more complicated flow field, particularly when a ceiling is present. As indicated in Fig. 10, fires in a corner rise vertically in the corner until the gases impinge on the ceiling, at which point the fire will be redirected along the ceiling and the top of the walls. Near the top of the walls, flaming vortices will flow out from the corner resulting in elevated heat fluxes along the top of the wall as much as twice the ceiling jet thickness.

Incident heat flux correlations in a corner with a ceiling were developed by Lattimer et al. [91] The study was conducted using a 2.4 m high open corner

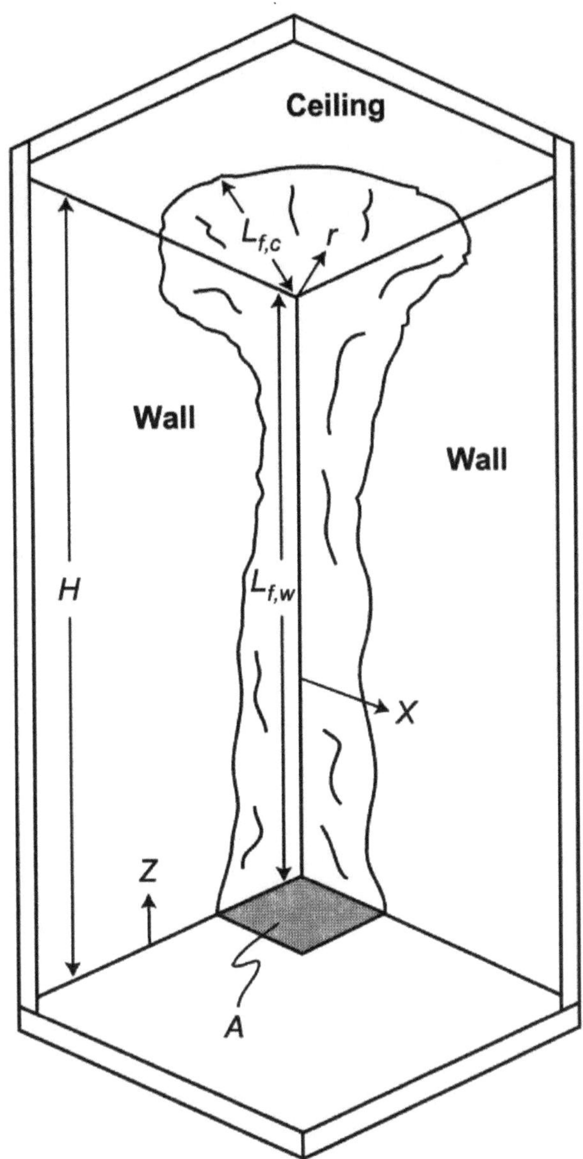

Fig. 10 Fire in a corner configuration

constructed of two walls and a ceiling. Fires were produced using square propane burners having single side lengths of 0.17, 0.30, and 0.50 m. Fires were placed directly in the corner. The study included fires with heat release rates ranging from 50 to 300 kW.

Correlations were developed for the three regions in the corner shown in Fig. 11. The regions were the corner walls on the lower part of the walls, the top portion of

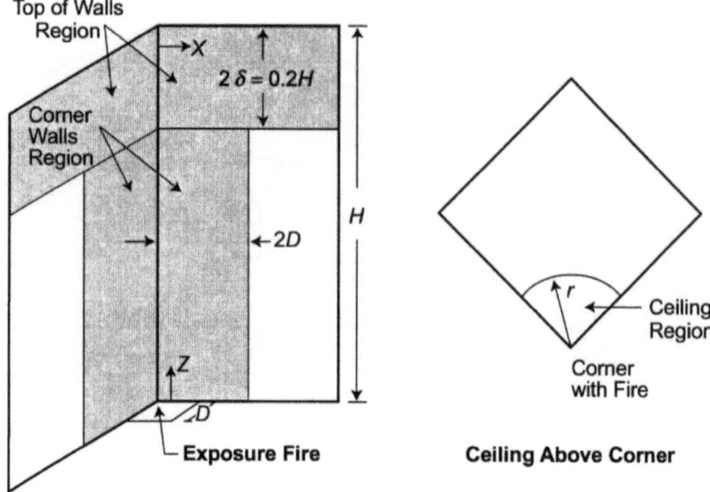

Fig. 11 Corner with a ceiling configuration showing the three regions where incident heat flux correlations were developed in the study of Lattimer et al. [91]

the walls near the ceiling, and along the ceiling. The corner walls region extended from the fire to approximately 1.8 m above the floor. Above this region, the incident heat flux onto the walls was measured to be affected by the hot gases flowing along the ceiling. The distance of 1.8 m is approximately twice the ceiling jet thickness below the ceiling or $H - 2\delta$ where $H = 2.2$ m and $\delta = 0.1H$ [92]. Correlations for the top part of the walls, which are heated by the ceiling jet, were developed using data from 1.8 m to 2.2 m or $H - 2\delta < z < H$.

The flame length in the corner with a ceiling was taken to be the flame length in the corner plus any flame extension along the ceiling. The following relation can be used to calculate the flame tip length with the fire in the corner:

$$L_{f,\text{tip}} = 5.9 Q_D^{*1/2} D \quad \text{(m)} \tag{23}$$

where:

$$Q_D^* = \frac{\dot{Q}}{\rho_0 c_p T_0 \sqrt{g} D^{5/2}} \quad (-) \tag{24}$$

$\dot{Q} = $ Heat release rate of the fire (kW)
$D = $ Diameter of the fuel package (m)
$\rho_0 = $ Density of air at initial ambient conditions (1.2 kg/m^3)
$c_p = $ Specific heat capacity of air at initial ambient conditions [1.0 kJ/(kg K)]
$T_o = $ Temperature at initial ambient conditions (293 K)
$g = $ Gravitational acceleration (9.81 m/s^2)

Flame lengths are taken relative to the base of the fire. This correlation can be used to estimate flame lengths in a corner with or without a ceiling. For noncircular fuel packages with a length-to-width ratio of near one, the equivalent diameter of the fuel package can be estimated using the surface area, A, of the noncircular fuel package:

$$D = \sqrt{\frac{4A}{\pi}} \quad (\text{m}) \tag{25}$$

where:

A = Surface area of the fuel package (m^2)

Walls at Corner

Correlations in this section can be used to estimate the incident heat flux in the corner with the fire. These correlations can be used to estimate the incident heat flux in a corner configuration with or without a ceiling. When a ceiling is present, the correlations are valid up to an elevation of $z = H - 2\delta$, where $\delta = 0.1H$ [92]. Along the height of the walls in the corner, the peak heat fluxes were typically measured near the base of the fire. These peak heat fluxes were measured to be a function of the fire diameter as shown in Fig. 12. The curve in Fig. 12 is a correlation to the data and is expressed using Eq. 26:

$$\dot{q}''_{\text{peak}} = 120[1 - \exp(-4.0D)] \quad (\text{kW}/\text{m}^2) \tag{26}$$

where:

$\dot{q}''_{\text{peak}}$ = Peak heat flux in the corner (kW/m^2)
D = Diameter of the fuel package (m)

Fig. 12 Peak heat flux along the height of the walls in the corner. (Data from Lattimer et al. [91])

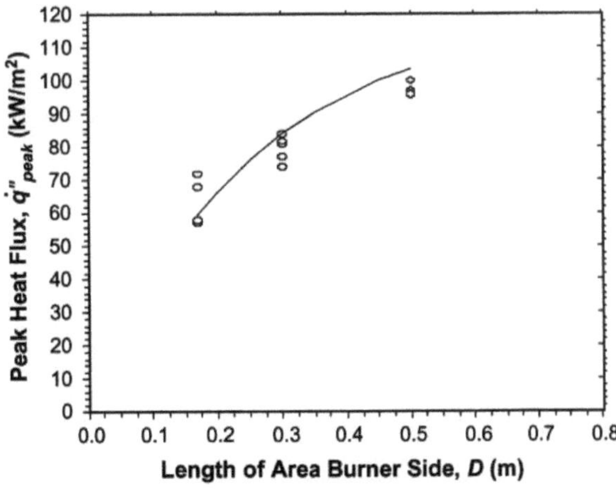

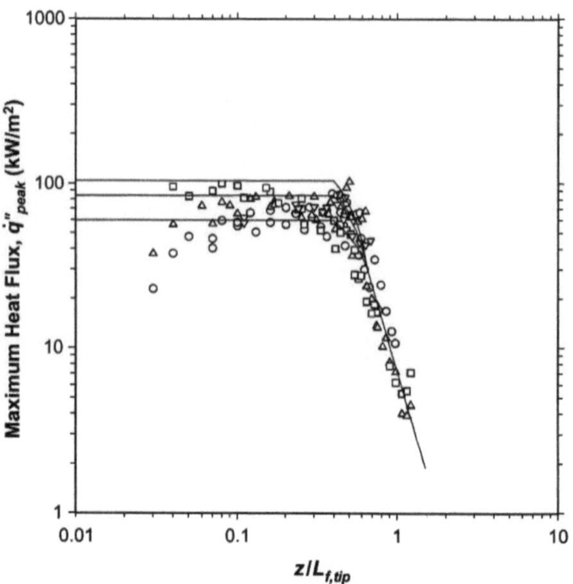

Fig. 13 Maximum heat fluxes to the walls near the corner with square burner sides of ○-0.17 m, △-0.30 m, ▽-0.30 m (elevated), and □-0.50 m and fire sizes from 50 to 300 kW. (Data from Lattimer et al. [91])

The vertical distribution in the maximum heat flux along the walls near the corner is shown in Fig. 13 plotted with the elevation above the fire, z, normalized with respect to the flame tip length. Peak heat flux levels were measured in the lower part of the flame ($z/L_{f,tip} \leq 0.4$) and decreased with distance above $z/L_{f,tip} = 0.4$. A general correlation to represent this behavior is as follows:

$$\dot{q}''_{max} = \dot{q}''_{peak} \quad z/L_f \leq 0.4 \quad (kW/m^2) \tag{27a}$$

$$\dot{q}''_{max} = \dot{q}''_{peak} - 4\left[\frac{z}{L_{f,tip}} - \frac{2}{5}\right]\left(\dot{q}''_{peak} - 30\right) \quad 0.4 \leq z/L_f \leq 0.65 \quad (kW/m^2) \tag{27b}$$

$$\dot{q}''_{max} = 7.2\left(\frac{z}{L_{f,tip}}\right)^{-10/3} \quad z/L_f \geq 0.65 \quad (kW/m^2) \tag{27c}$$

where:

$\dot{q}''_{max}$ = Maximum heat flux at a particular elevation in the corner (kW/m^2)
$\dot{q}''_{peak}$ = Peak heat flux in the corner (kW/m^2)
z = Elevation along the flame height in the corner (m)
$L_{f,\,tip}$ = Flame tip length calculated using Eqs. 23 and 24 (m)

Heat fluxes will decay with distance away from the corner as shown in Fig. 14. Significant heat fluxes can exist as far as two fire diameters horizontally out from the corner. For a conservative analysis, the maximum vertical heat flux distribution measured in the corner should be assumed from the corner to two fire diameters horizontally out from the corner.

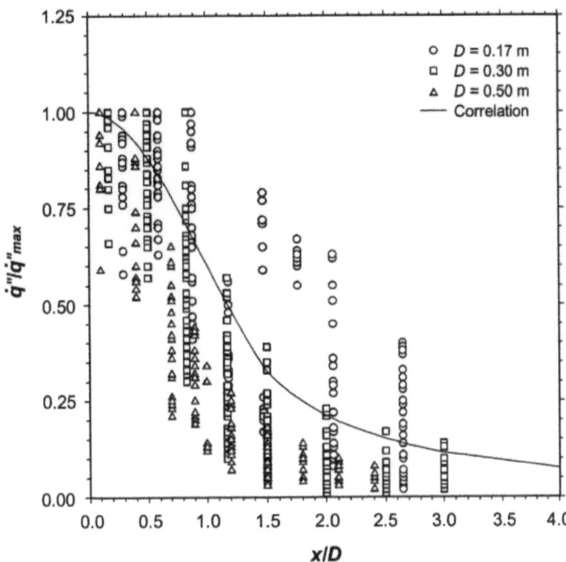

Fig. 14 Heat flux distribution horizontally out from the corner on the lower part of the corner walls

Top of Walls

This section provides correlations to estimate heat fluxes along the top of the walls in a corner configuration with a ceiling. These incident heat flux correlations apply to the top of the walls approximately twice the ceiling jet thickness below the ceiling *or* $H - 2\delta < z < H$ where $\delta = 0.1H$ [92]. Along the top part of the wall, the maximum heat fluxes were measured at locations less than 0.15 m below the ceiling. The maximum heat fluxes are shown in Fig. 15 plotted against the dimensionless distance along the flame, $(x + H)/L_{f,tip}$. These heat fluxes can be estimated using the following relations:

$$\dot{q}''_{max} = 120 \quad (x + H)/L_{f,tip} \leq 0.58 \quad (\text{kW/m}^2) \tag{28a}$$

$$\dot{q}''_{max} = 18\left[(x + H)/L_{f,tip}\right]^{-3.5} \quad (x + H)/L_{f,tip} > 0.58 \quad (\text{kW/m}^2) \tag{28b}$$

where:

$x =$ Distance horizontally out from the corner (m)
$H =$ Distance between the base of the fire and the ceiling (m)
$L_{f,tip} =$ Flame tip length calculated using Eqs. 23 and 24 (m)

The assumed plateau in the correlation was based on the maximum heat flux levels measured in larger fire tests with burning boundaries [91]. Heat fluxes will decrease with distance below the ceiling. Conservatively, it can be assumed that incident flux along the top of the walls is constant with distance below the ceiling and is equal to the maximum incident flux predicted through Eqs. 28a and 28b.

Fig. 15 Maximum heat flux along the top of the walls during corner fire tests with square burner sides of ○-0.17 m, **Δ-0.30 m**, **∇-0.30** m (elevated), and □-**0.50** m and fire sizes from 50 **to 300 kW.** (Data from Lattimer et al. [91])

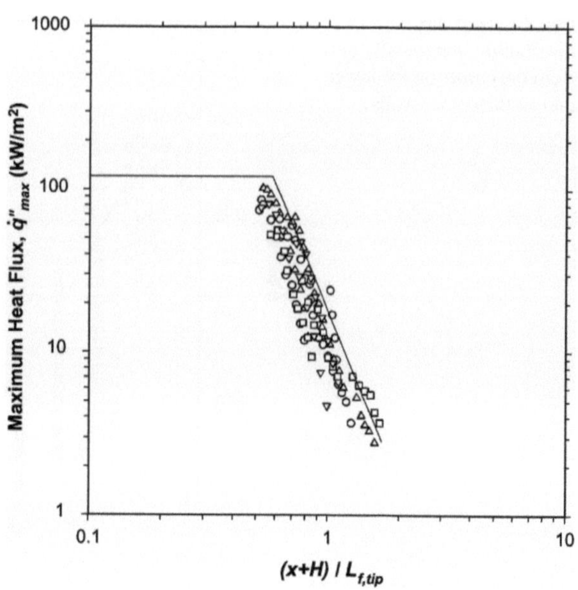

Ceiling Above a Corner

Correlations in this section can be used to predict the incident heat flux distribution radially out from a corner along the ceiling. The heat fluxes to the ceiling were determined to be a function of dimensionless distance along the flame length, $(r + H)/L_{f,\text{tip}}$. A plot of the heat fluxes measured along the ceiling out from the corner is shown in Fig. 16. A correlation to predict the heat flux distribution along the ceiling is as follows:

$$\dot{q}'' = 120 \quad (r + H)/L_{f,tip} \leq 0.58 \quad (\text{kW/m}^2) \tag{29a}$$

$$\dot{q}'' = 18\left[(r + H)/L_{f,\text{tip}}\right]^{-3.5} \quad (r + H)/L_{f,\text{tip}} > 0.58 \quad (\text{kW/m}^2) \tag{29b}$$

where:

$r =$ Radial distance from the corner (m)
$H =$ Distance between the base of the fire and the ceiling (m)
$L_{f,\text{tip}} =$ Flame tip length calculated using Eqs. 23 and 24 (m)

This correlation is similar to the one developed for predicting the maximum heat flux along the top of the walls, Eqs. 28a and 28b, except the length scale here is r instead of x. Again, the assumed plateau in the correlation was based on the maximum heat flux levels measured in larger fire tests with burning boundaries [91]. The heat flux at the impingement point can be estimated using Eqs. 29a and 29b with $r = 0$.

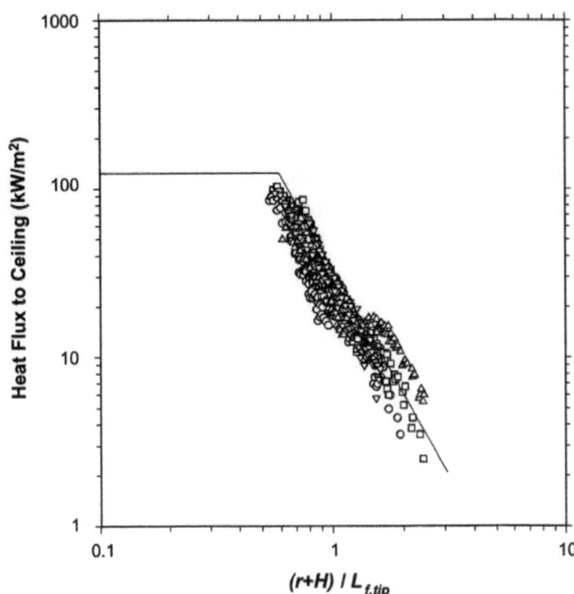

Fig. 16 Heat flux along the ceiling above a fire in a corner during tests with square burner sides of ○-0.17 m, ▲-0.30 m, ▽-0.30 m (elevated), and □-0.50 m and fire sizes from 50 to 300 kW. (Data from Lattimer et al. [91])

Data Requirements

1. Diameter of the fuel package, D. For noncircular fuel packages, the equivalent diameter may be calculated using Eq. 25 and the surface area of the fuel package.
2. Heat release rate of the fire, $\dot{Q}$.
3. Distance between the base of the fire and the ceiling, H.
4. Location along the surface where incident heat flux level is needed. This could be the elevation along the height of the corner, z, horizontal distance from the corner along the top of the walls, x, or radially out from the corner along the ceiling, r.

Data Sources

1. Heat release rate data may be obtained from Babrauskas [69], Hoglander and Sundstrum [70], or Mudan and Croce. [71]

Assumptions

This analysis assumes that the fire is attached to the corner walls, the corner walls are vertical and at a 90° angle, and the ceiling is horizontal and at a 90° angle with the corner walls. Walls that are not vertical may result in different total incident heat flux levels as a result of the flame's becoming separated from the wall or the difference in entrainment into the plume. Incident heat fluxes to the corner walls across the width of the fire are constant and are equal to the maximum vertical heat flux distribution in the corner. Heat fluxes along the top of the walls are constant and equal to the maximum horizontal heat flux distribution along the top of the walls.

Validation

Other studies have been conducted with propane fires in a corner configuration with and without a ceiling. Corner heat flux data with no ceiling [93] agree well with the heat flux data in Fig. 13 when considered relative to the flame tip. In a study with fires in a corner and a ceiling, Hasemi et al. [94] measured incident heat flux levels on the walls and ceiling from both exposure fires and simulated burning boundaries. Trends in incident heat flux levels measured by Hasemi et al. [94] along the top of the walls and the ceiling agree well with the data in Figs. 15 and 16 when using the dimensionless distances used in these figures. In tests with propane gas burners against a non-combustible boundary, similar heat flux levels have been measured by other investigators for limited conditions [87, 88]. Ohlemiller, Cleary, and Shields [95] measured peak heat fluxes approximately 10% to 20% higher using similar size propane square burners.

Lattimer et al. [91] also demonstrated that the correlations for incident heat fluxes in the three regions of the corner configuration also hold when the boundary is combustible and burning. For this case, a modified length scale is required to correctly predict flame length.

Limitations

Correlations for incident heat fluxes were developed using luminous flames in an open environment with the fire directly in the corner. Using these relations inherently assumes:

- There is negligible heating from a hot gas layer in the surroundings.
- The fire is against the wall.
- The flames are luminous.
- The corner walls are vertical and at a 90° angle.
- The ceiling is horizontal and at a 90° angle with the corner walls.

The experimental study considered fire diameters as large as 0.50 m and heat release rates as large as 300 kW. No data were available to validate the correlations against fires with larger diameters or higher heat release rates. The presence of a hot gas layer may increase the total incident flux onto the wall, and if significant in the area of interest, adding this contribution to the total incident heat flux from the fire plume may be warranted [89]. Moving the fire away from the corner will eventually cause the incident heat fluxes to become lower, largely because the flame becomes detached from the wall [90]. Thus, the use of correlations in this section for fires that may be slightly spaced from the corner will yield conservative results. Flames less luminous than those produced by the propane fires (i.e., natural gas) may transmit lower total incident heat fluxes to the surfaces because the radiative heat flux to the wall will be lower [87, 88, 96]. The propane fuel fires used to develop the heat flux data presented in this section produce a moderate amount of soot; therefore, the heat flux levels presented in this section should be considered to be average but not bounding for all different fuels. Propane burners are also used extensively in standard fire tests as an exposure fire that is representative of real fires. Therefore, the incident heat fluxes from these flames are considered to be representative of those produced by most fires.

Fires Impinging on Unbounded Ceilings

Fires that impinge onto an unbounded ceiling as shown in Fig. 17 have flames that are redirected radially out from the impingement point. The highest heat fluxes onto the ceiling will be at the impingement or stagnation point. Heat fluxes will tend to decrease with radial distance away from the stagnation point. Correlations are provided in this section to estimate the heat fluxes from such a fire to the ceiling.

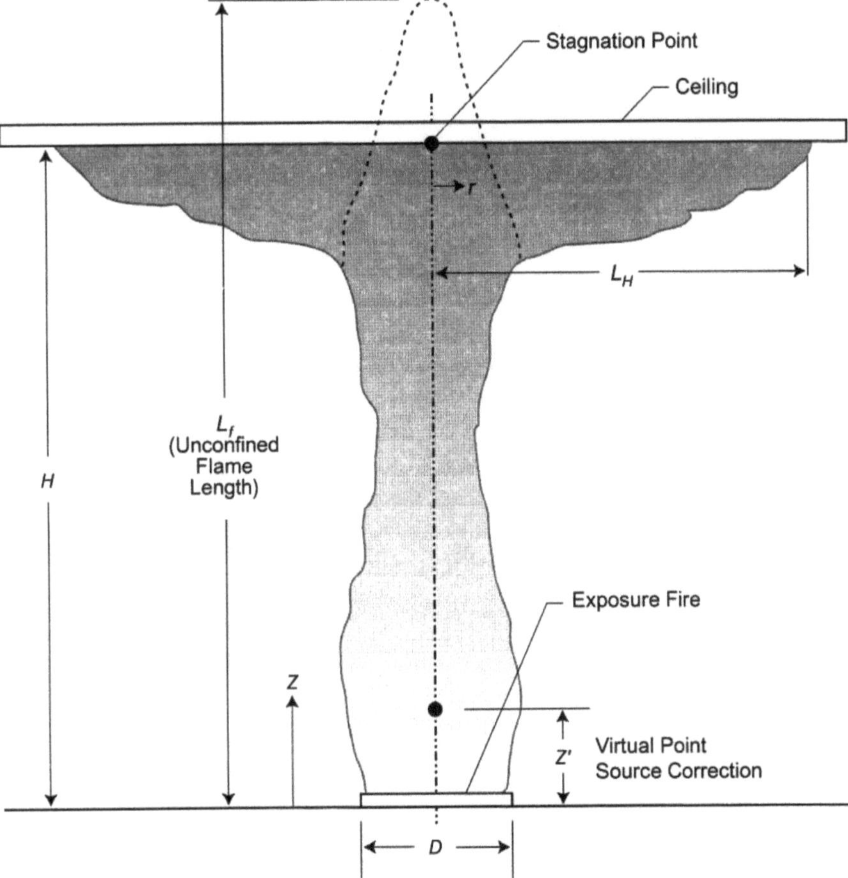

Fig. 17 Unbounded ceiling configuration

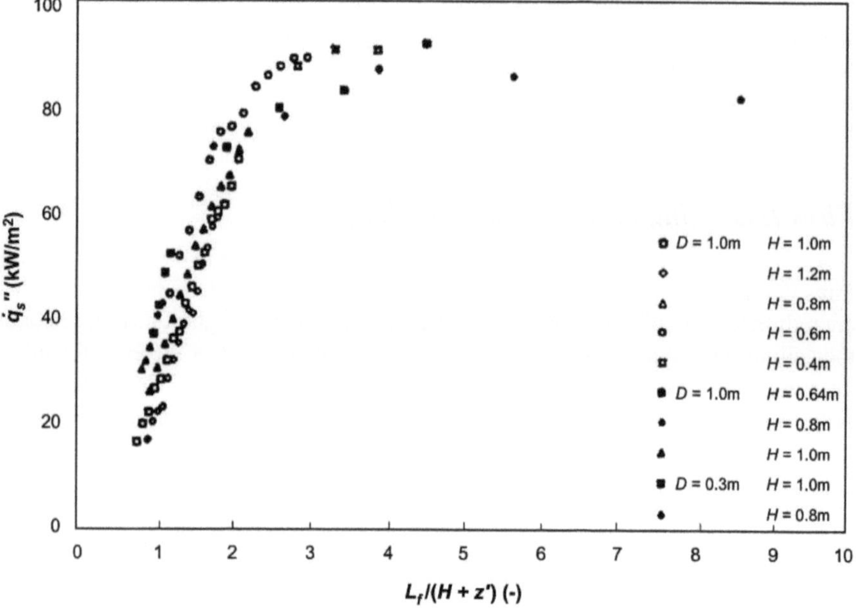

Fig. 18 Stagnation point heat fluxes on an unbounded ceiling with a fire impinging on it. (Data from Hasemi et al. [97])

The incident heat flux due to a fire impinging onto an unbounded flat ceiling has been experimentally characterized by Hasemi et al. [97] In this study, Hasemi et al. [97] conducted a series of fire tests using propane gas burners located at different distances beneath a non-combustible unbounded ceiling. The test configuration is shown in Fig. 17 along with important variables. Fires as large as approximately 400 kW were considered in the study. Heat flux gauges were used to measure the incident heat flux along the ceiling both directly above the centerline of the fire (i.e., stagnation point) and radially out from the stagnation point.

A plot of the heat flux levels at the stagnation point is shown in Fig. 18. Heat fluxes at the stagnation point are shown in this figure to plateau at approximately 90 kW/m². In order to collapse the data, the unconfined flame tip length was normalized with respect to the distance between the ceiling and fire, H, plus the virtual source origin correction, z'. The unconfined fire flame tip length was calculated using the following relation:

$$L_f = 3.5 Q_D^{*n} D \quad \text{(m)} \tag{30}$$

where:

$n = 2/5$ for $Q^*_D > 1.0$

$n = 2/3$ for $Q^*_D < 1.0$

$$Q_D^* = \frac{\dot{Q}}{\rho_0 c_p T_0 \sqrt{g} D^{5/2}} \tag{31}$$

$\dot{Q}$ = Heat release rate of the fire (kW)
D = Diameter of the fuel package (m)
ρ_0 = Density of air at initial ambient conditions (1.2 kg/m^3)
c_p = Specific heat capacity of air at initial ambient conditions [1.0 kJ/(kg K)]
T_0 = Temperature at initial ambient conditions (293 K)
g = Gravitational acceleration (9.81 m/s^2)

For noncircular fuel packages with a length-to-width ratio of near one, the equivalent diameter of the fuel package can be estimated using the surface area, A, of the noncircular fuel package:

$$D = \sqrt{\frac{4A}{\pi}} \quad (\text{m}) \tag{32}$$

where:

A = Surface area of the fuel package (m^2)

The virtual point source correction for this geometry was determined using the following relations:

$$z' = 2.4D\left(Q_D^{*2/5} - Q_D^{*2/3}\right) \quad Q_D^* < 1.0 \quad (\text{m}) \tag{33a}$$

$$z' = 2.4D\left(1 - Q_D^{*2/5}\right) \quad Q_D^* \geq 1.0 \quad (\text{m}) \tag{33b}$$

where:

Q^*_D = Dimensionless quantity defined in Eq. 31
D = Diameter of the fuel package (m)

The radial distribution in the incident heat flux decays with distance from the stagnation point as shown in Fig. 19. The length of the flame used to correlate this data was the measured flame extension plus a virtual origin correction. The measured flame extension was defined as the distance between the fire and the ceiling, H, plus the radial extension of the flame out from the center of the fire, L_H. The location of the flame tip in this geometry was found to correlate with Q^*_H, which is defined the same as in Eq. 31 except the length scale is H instead of D. The flame tip correlation was determined to be:

$$L_{f,\text{tip}} = (L_H + H) = 2.89 Q_H^{*1/3} H \quad (\text{m}) \tag{34}$$

where:

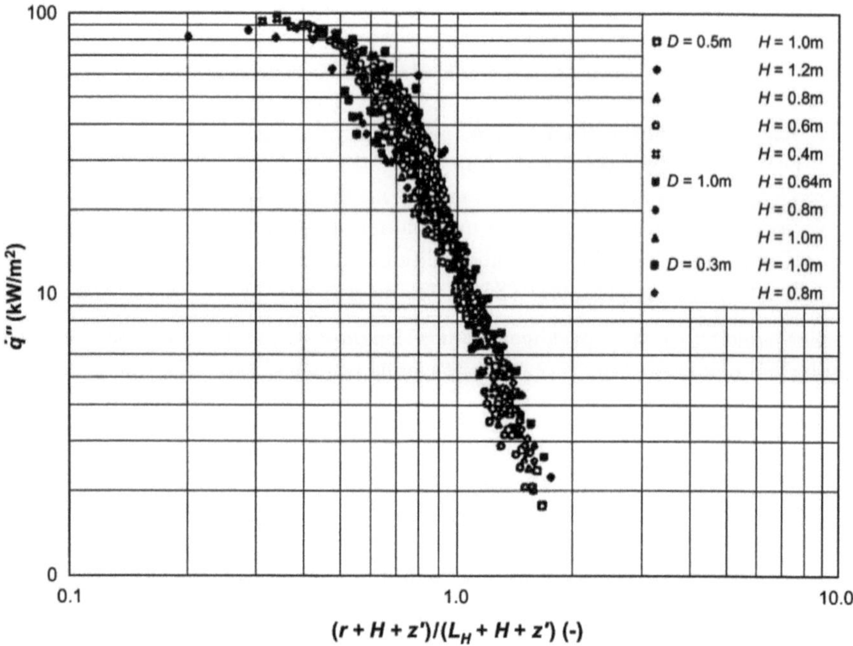

Fig. 19 Heat fluxes to a ceiling due to a propane fire impinging on the surface. (Data from Hasemi et al. [97])

$$Q_H^* = \frac{\dot{Q}}{\rho_0 c_p T_0 \sqrt{g} H^{5/2}} \quad (-) \tag{35}$$

L_H = Flame extension along the ceiling from the stagnation point to the flame tip (m)
H = Distance between the base of the fire and the ceiling (m)
$\dot{Q}$ = Heat release rate of the fire (kW)
ρ_0 = Density of air at initial ambient conditions (1.2 kg/m³)
c_p = Specific heat capacity of air at initial ambient conditions [1.0 kJ/(kg K)]
T_0 = Temperature at initial ambient conditions (293 K)
g = Gravitational acceleration (9.81 m/s²)

The radial heat flux distribution along the ceiling at $w > 0.45$ can be estimated using the correlation recommended by Wakamatsu [98]:

$$\dot{q}'' = 518.8 e^{-3.7w} \quad w > 0.45 \quad (kW/m^2) \tag{36a}$$

where:

$$w = (r + H + z')/(L_H + H + z') \quad (-) \tag{36b}$$

r = Radial distance along the ceiling from the stagnation point (m)

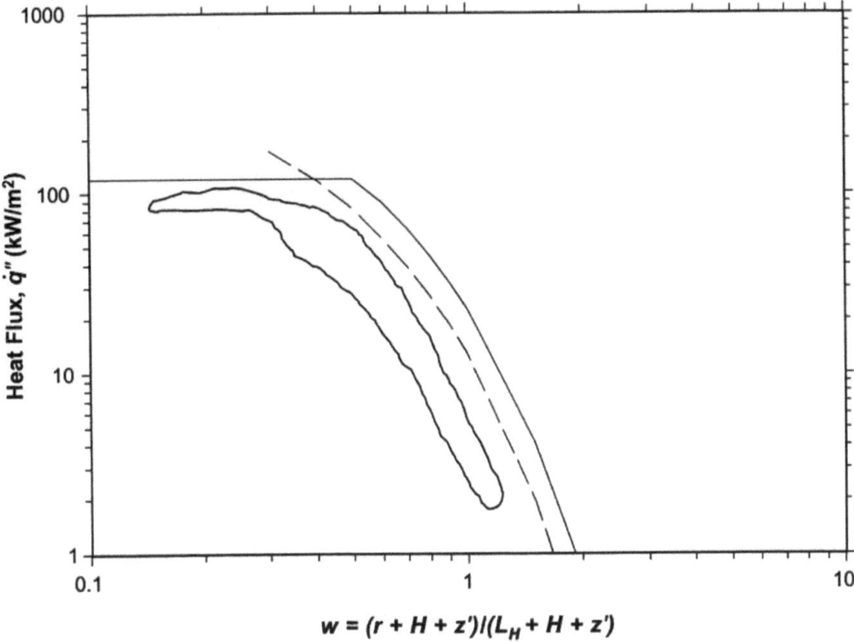

Fig. 20 Comparison of the best fit curve proposed by Wakamatsu (– –) and a bounding fit to the data (—). The unbounded ceiling data of Hasemi et al. [97] is represented as the outlined area

H = Distance between the base of the fire and the ceiling (m)
z' = Virtual source origin correction (m)
L_H = Flame extension along the ceiling from the stagnation point to the flame tip (m)

Figure 20 contains a plot of Eqs. 36a and 36b (dashed line) along with a representation of the data of Hasemi et al. [97] for a flat unbounded ceiling. As noted in Eqs. 36a and 36b, this correlation adequately estimates the data when $w > 0.45$ but significantly overestimates heat flux levels for smaller values of w. Based on the data from Hasemi et al. [97] and other data from fires impinging on I-beams mounted to a ceiling [98], a correlation was developed to predict the bounding heat flux levels where w is defined in Eq. 36b:

$$\dot{q}'' = 120 \quad w \leq 0.5 \quad (\text{kW/m}^2) \tag{37a}$$

$$\dot{q}'' = 682 \exp(-3.4w) \quad w > 0.5 \quad (\text{kW/m}^2) \tag{37b}$$

This correlation is shown in Fig. 20 as the solid line. The peak heat flux of 120 kW/m² at $w \leq 0.5$ bounds nearly all the heat flux measurements made in this range for the studies of Hasemi et al. [97] and Myllymaki and Kokkala [98].

Data Requirements

1. Diameter of the fuel package, D. For noncircular fuel packages, the equivalent diameter may be calculated using Eq. 32 and the surface area of the fuel package.
2. Heat release rate of the fire, $\dot{Q}$.
3. Distance between the base of the fire and the ceiling, H.
4. Radial location out from the centerline of the fire, r, where the incident heat flux level is needed.

Data Sources

1. Heat release rate data may be obtained from Babrauskas [69], Hoglander and Sundstrum [70], or Mudan and Croce. [70]

Assumptions

The fire is assumed to be impinging on a horizontal, flat ceiling far from walls or any other obstructions.

Validation

Several experimental and theoretical studies have been performed on fires impinging on an unbounded *ceiling* [77–80, 97, 99–101]. Total heat fluxes from fires and fire plumes impinging on the ceiling were measured by Hasemi et al. [97], You and Faeth [79, 80], and Koltkala [77, 78]. Due to the fuel type and size of fires evaluated, heat flux levels measured by Hasemi et al. [97] were higher than those measured in other studies. Therefore, the correlations developed using the data of Hasemi et al. are considered conservative.

Limitations

Correlations for incident heat fluxes were developed using luminous flames in an open environment with the fire beneath an unbounded flat ceiling. Using these relations inherently assumes negligible heating from a hot gas layer in the surroundings, the flames are luminous, and the ceiling is horizontal. The presence of a hot gas layer may increase the total incident flux onto the wall, and if significant in the area of interest, adding this contribution to the total incident heat flux from the fire plume may be warranted. Flames less luminous than those produced by the propane fires (i.e., natural gas) may transmit lower total incident heat fluxes to the wall because the radiative heat flux to the wall will be lower. Propane flames do not have the highest soot production of any fuel, and, therefore, incident heat fluxes may not be bounding. However, propane burners are used extensively in standard fire tests as an exposure fire that is representative of real fires. Therefore, the incident heat fluxes from these flames are considered to be representative of those produced by most fires.

Fire Impinging on a Horizontal I-Beam Mounted Below a Ceiling

The final geometry considered is an I-beam that is mounted below a ceiling as shown in Fig. 21, with the fire impinging on the lower flange of the I-beam. The focus here is the heat fluxes from the fire onto the I-beam. This case turns out to be quite similar to a fire impinging onto an unbounded ceiling.

Two separate studies have been conducted to evaluate the heat flux incident onto an I-beam mounted below a ceiling with an exposure fire impinging upon the beam (Hasemi et al. [97], Wakamatsu et al. [102], and Myllymaki and Kokkala [98]). In these studies, the heat flux was measured along the four surfaces of the I-beam noted in Fig. 21:

1. Downward face of the lower flange
2. Upward face of the lower flange
3. The web
4. Downward face of the upper flange

The I-beam evaluated in these studies was 3.6 m long, a web 150 mm high and 5 mm thick, and flanges 75 mm wide and 6 mm thick. For each of these surfaces, heat fluxes were measured from the stagnation point of the fire (centerline of the fire) along the length of the I-beam.

Results from these studies have demonstrated that the incident heat flux onto all surfaces of the beam will be equal to or less than the heat flux levels measured with a fire impinging onto a flat unbounded ceiling. Wakamatsu et al. [102] measured this for fires up to 900 kW. Flame lengths were observed to be different along the lower flange, upper flange, and center of the web of the I-beam. Correlations to predict these flame lengths were developed for the lower flange [102],

$$L_{f,\text{tip}B} = (L_B + H_B) = 2.3 Q_{H_B}^{*0.3} H_B \tag{38}$$

where:

$$\dot{Q}_{H_B} = \frac{\dot{Q}}{\rho_0 c_p T_0 \sqrt{g H_B}^{5/2}} \quad (-) \tag{39}$$

the upper flange [102],

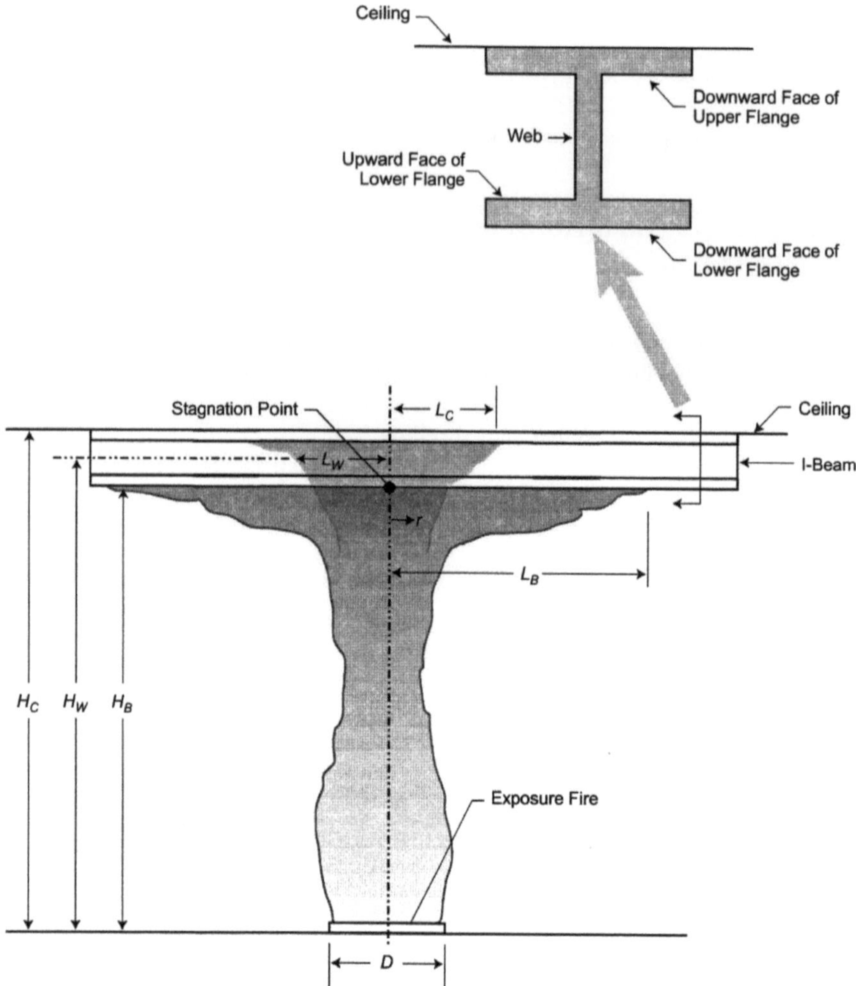

Fig. 21 I-beam mounted below an unbounded ceiling

L_B = Flame extension along the lower flange from the stagnation point to the flame tip (m)

L_C = Flame extension along the upper flange from the stagnation point to the flame tip (m)

L_W = Flame extension along the web center from the stagnation point to the flame tip (m)

H_B = Distance between the base of the fire and bottom of the lower flange (m)

H_C = Distance between the base of the fire and the ceiling (m)

H_W = Distance between the base of the fire and the center of the web (m)

$\dot{Q}$ = Heat release rate of the fire (kW)

ρ_0 = Density of air at initial ambient conditions (1.2 kg/m^3)

c_p = Specific heat capacity of air at initial ambient conditions [1.0 kJ/(kg K)]

T_0 = Temperature at initial ambient conditions (293 K)

g = Gravitational acceleration (9.81 m/s^2)

$$L_{f,tipC} = (L_C + H_C) = 2.9 Q_{H_C}^{*0.4} H_C \tag{40}$$

where:

$$Q_{H_C}^* = \frac{\dot{Q}}{\rho_0 c_p T_0 \sqrt{g} H_C^{5/2}} \quad (-) \tag{41}$$

and for the center of the web [98],

$$L_{f,tipW} = (L_W + H_W) = 2.9 Q_{H_W}^{*0.4} H_W \quad (m) \tag{42}$$

where:

$$Q_{H_W}^* = \frac{\dot{Q}}{\rho_0 c_p T_0 \sqrt{g} H_W^{5/2}} \quad (-) \tag{43}$$

The form of these correlations is similar to that of the unbounded ceiling flame length correlation given in Eq. 34. The dimensionless distance along the flame beneath the downward face of the lower flange was taken to be

$$w = (r + H_B + z')/(L_B + H_B + z') \quad (-) \tag{44}$$

where:

$r = $ Radial distance along the I-beam from the stagnation point (m)
$H_B = $ Distance between the base of the fire and the lower flange (m)
$z' = $ Virtual source origin correction (m)
$L_B = $ Flame extension along the lower flange from the stagnation point to the flame
tip (m)

The dimensionless distance for the upper flange on the I-beam was taken to be

$$w = (r + H_C + z')/(L_C + H_C + z') \quad (-) \tag{45}$$

where:

$r = $ Radial distance along the I-beam from the stagnation point (m)
$H_C = $ Distance between the base of the fire and the upper flange (m)
$z' = $ Virtual source origin correction (m)
$L_C = $ Flame extension along the upper flange from the stagnation point to the flame
tip (m)

The dimensionless distance for the web on the I-beam was taken to be

$$w = (r + H_W + z')/(L_W + H_W + z') \quad (-) \tag{46}$$

where:

r = Radial distance along the I-beam from the stagnation point (m)
H_W = Distance between the base of the fire and the center of the web (m)
z' = Virtual source origin correction (m)
L_W = Flame extension along the web center from the stagnation point to the flame tip (m)

The incident heat flux levels measured by Wakamatsu et al. [102] on the different faces of the I-beam are shown in Fig. 22. On the downward face of the lower flange (where the fire was directly impinging), heat flux levels along the flame length were measured to be similar to the incident heat fluxes measured along a flame under an unbounded ceiling. However, all other surfaces of the I-beam had heat fluxes somewhat lower than those measured along a flame under an unbounded ceiling.

The study of Myllymaki and Kokkala [98] considered the effects of larger fires (up to 3.9 MW) on the heat flux incident on the different faces of the I-beam. Some of the heat flux measurements made in this study are shown in Fig. 23. In this study, Myllymaki and Kokkala [98] found that, for fires over 2.0 MW, the incident heat fluxes onto all faces of the I-beam were equivalent to or slightly higher than those measured along an unbounded ceiling.

Data from these studies demonstrate that the heat flux to the I-beam can be conservatively estimated using the bounding heat flux correlation in Eqs. 47a and 47b using the appropriate expression for w provided in Eqs. 44 through 46:

$$\dot{q}'' = 120 \quad w \leq 0.5 \quad (\text{kW/m}^2) \tag{47a}$$

$$\dot{q}'' = 682 \exp(-3.4w) \quad w > 0.5 \quad (\text{kW/m}^2) \tag{47b}$$

Data Requirements

1. Diameter of the fuel package, D. For noncircular fuel packages, the equivalent diameter may be calculated using Eq. 32 and the surface area of the fuel package.
2. Heat release rate of the fire, $\dot{Q}$.
3. Distance between the base of the fire and the bottom flange, the center of the web, and the top of the flange.
4. Distance out from the impingement point on the I-beam where the heat flux is needed, r.

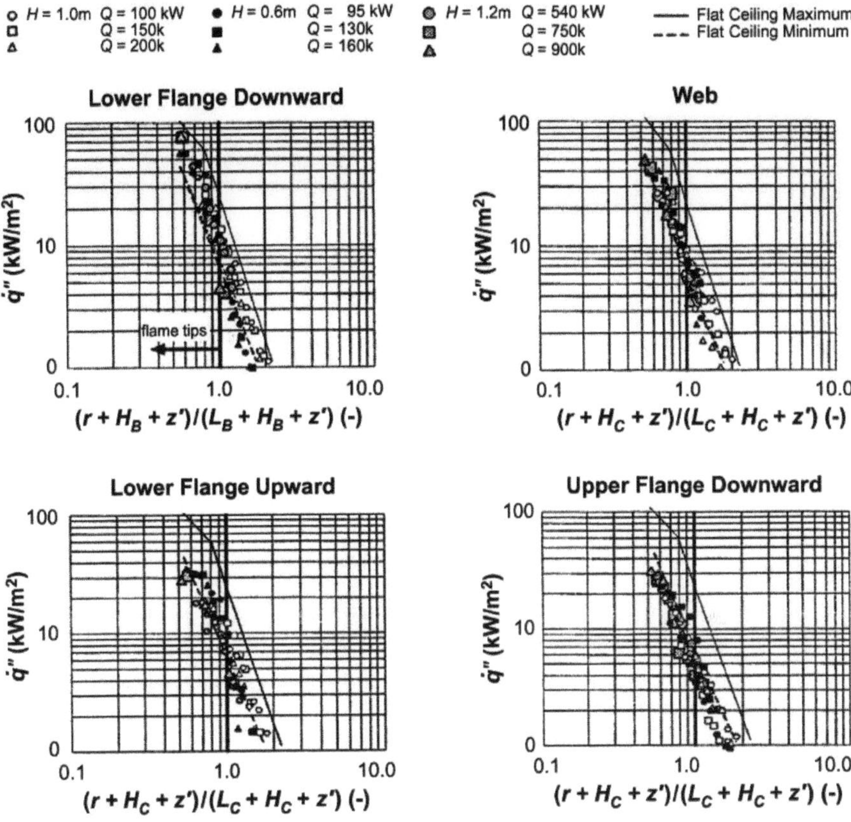

Fig. 22 Heat flux measured onto the surfaces of an I-beam mounted below an unbounded ceiling for fires 95 to 900 kW [102]

Data Sources

1. Heat release rate data may be obtained from Babrauskas [69], Hoglander and Sundstrum [70], or Mudan and Croce [71].

Assumptions

The I-beam being analyzed should have similar dimensions to the one considered in these two studies (3.6 m long, a web 150 mm high and 5 mm thick, and flanges 75 mm wide and 6 mm thick), and the fire is assumed to be impinging directly onto the bottom flange of the I-beam. The I-beam is also assumed to be located remote from any walls or ceiling obstructions.

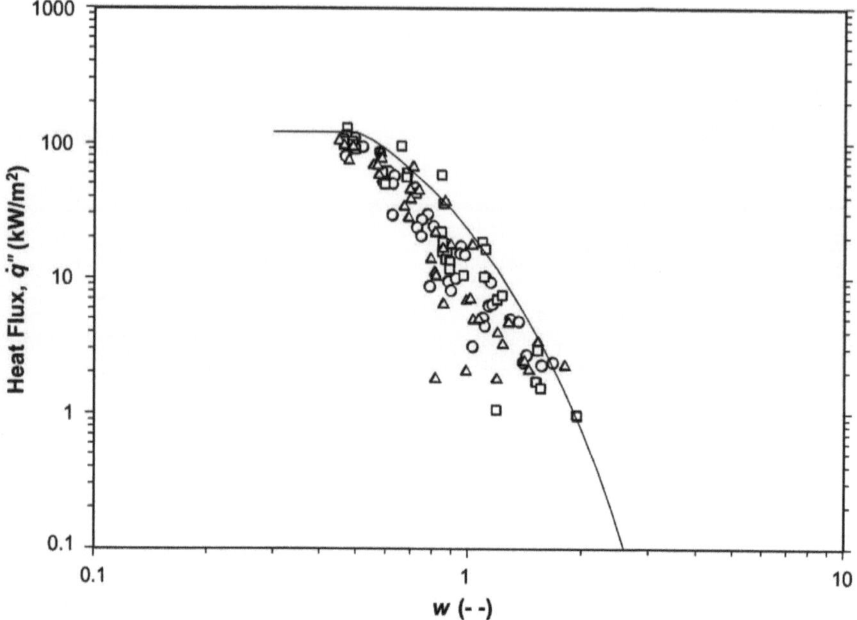

Fig. 23 Heat flux measured on the ○-bottom flange, □-web, and △-upper flange of an I-beam mounted below an unbounded ceiling for fires 565–3870 kW [98]. The line in the plot is the curve given in Eqs. 47a and 47b

Validation

These two studies provide a good validation of the heat fluxes experienced by the particular I-beam tested. Results produced using propane fuel fires agreed well with the larger liquid heptane pool fire tests.

Limitations

The height of the webbing and the width of the flanges may affect the heat fluxes to the I-beam. Other size I-beams have not been tested to evaluate the impact of I-beam dimensions on heat flux. Correlations for incident heat fluxes were developed using luminous flames in an open environment with the fire directly impinging on the I-beam. Using these relations inherently assumes negligible heating from a hot gas layer in the surroundings and that the I-beam is not located near any boundaries. The presence of a hot gas layer may increase the total incident flux onto the I-beam, and, if significant, this contribution should be added to the total incident heat flux from the fire plume [90]. Moving the fire away from the I-beam so that it does not impinge on the lower flange will change the heat flux distribution on the I-beam. These test data were developed with $0.48 < Q_H^* < 1.27$, fire distance below the lower flange of

$0.6 < H_B < 1.9$, fire diameters up to 1.6 m, and heat release rates up to 3.9 MW. Though results in this section indicate the heat flux is bounded by the correlation in Eqs. 47a and 47b, heat fluxes from large pool fires $(D > 1.6$ m$)$ impinging on an I-beam may be higher due to the changes in gas emissivity and flame temperature [67, 68].

Summary and Recommendations

The motivation for the work in the Bounding Heat Flux section has been the effect of the fire on objects in flames. Those studies were interested in the ability of nuclear waste casks or structural elements in offshore drilling facilities to withstand fire. On the other hand, the motivation for the work reported in the section on Heat Fluxes for Specific Geometries was primarily the effect of fire on ignition and fire growth (except for the I-beam studies). As a consequence, smaller exposure fires are considered in the latter section. For example, in the former, fires of up to 9 by 18 m were used (more than 300 MW) as compared to fires of up to 1 m at most or about 500 kW for the latter section. For the I-beam study, data include larger fires of 3.9 MW at most. The differences in the two sections are profound, and the reader should be aware of these distinctions in using the correlations. It is clear that pool-like fires exhibit higher temperatures and therefore higher heat fluxes as they become bigger. For example, a flame temperature of 1200 °C corresponds to a radiant heat flux of 267 kW/m^2. Yet in the Bounding section (Table 1), most measurements are more generally in the range of 150 kW/m^2, while for the smaller fires in the section on specific geometries, the upper limit of the heat flux measurements is more like 120 kW/m^2. Therefore, the user of this information must take into account the size and configuration of the fire. The type of fuel is less likely to be a factor.

Another issue that should be recognized in applying these results is that they are presented in terms of *incident* heat flux or the heat flux as measured to a cold target. In a design application, the heat flux that is *absorbed* into the structural element will decrease as the surface temperature increases. The boundary condition that should be used for the structure should account for the radiation loss for elements impacted by a fire plume:

$$-k\frac{\partial T}{\partial x} = \dot{q}'' - \varepsilon\sigma\left(T^4 - T_o^4\right)$$

where:

$\dot{q}'' =$ Incident heat flux given herein
$\varepsilon =$ Surface emissivity
$T_o =$ Cold target temperature

No factor of safety is addressed, and the user must be aware that that is not implicit in any of these results.

Appendices

Appendix A: Theoretical Examination of Methods

As can be seen in Fig. A.1, predictions of burning rate vary markedly among the different methods. Some of the methods assume stoichiometric or ventilation-limited burning, while others account for fuel-controlled burning.

Results by Harmathy for Wood Cribs

$$\dot{m}''_{F,\infty} = 6.2\,\text{g}/_{\text{m}^2\text{s}}, \quad \frac{\rho_0 \sqrt{g} A_o \sqrt{H_o}}{A_F} \geq 0.263$$

$$\dot{m}_F = 0.0236 \rho_0 \sqrt{g} A_o \sqrt{H_o}, \quad \frac{\rho_0 \sqrt{g} A_o \sqrt{H_o}}{A_F} < 0.263$$

From Eq. 20 (Chapter "Fully Developed Enclosure Fires"), similar results can be derived:

$$\frac{\dot{m}_F}{A_F} = \dot{m}''_{F,\infty}, \phi < 1 \ \text{ or } \ \frac{\rho_0 \sqrt{g} A_o \sqrt{H_o}}{A_F} \geq \frac{s \dot{m}''_{F,\infty}}{k_o} = \frac{4(0.0062)}{0.145} = 0.171$$

or

$$\dot{m}_F = \frac{k_o \rho_o \sqrt{g} A_o \sqrt{H_o}}{s} = \frac{0.145}{s} \rho_o \sqrt{g} A_o \sqrt{H_o} = 0.0363 \, \rho_o \sqrt{g} A_o \sqrt{H_o}$$

Fire Exposures to Structural Elements, The Society of Fire Protection Engineers Series,
https://doi.org/10.1007/978-3-031-64025-4

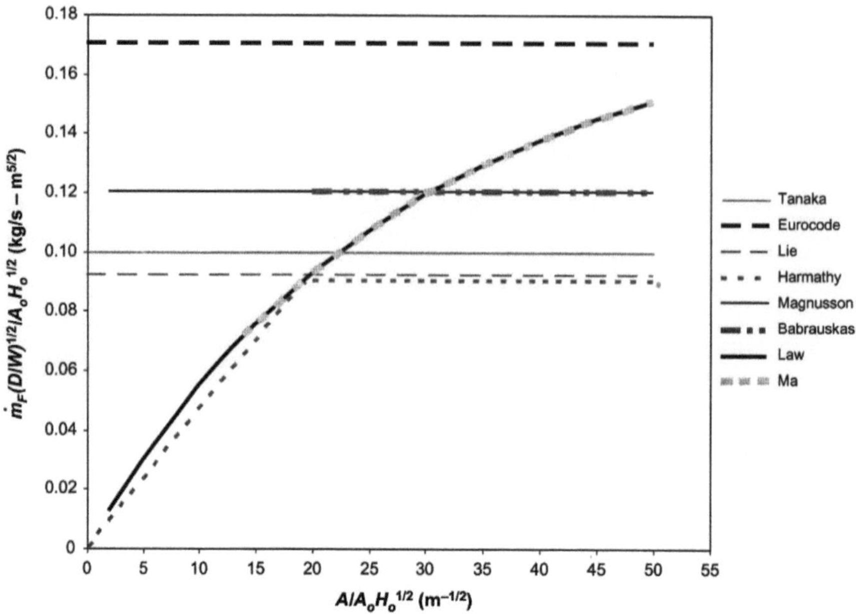

Fig. A.1 Comparison of burning rate predictions

Results by Bullen and Thomas for Pool Fires

$$\phi = \frac{\dot{m}_F s}{k_o \rho_o \sqrt{g} A_o \sqrt{H_o}}, \quad \begin{array}{l} s \approx 4 \text{ for wood,} \\ s \approx 7-10 \text{ for liquids.} \end{array}$$

The pool fire results are explained by the stoichiometry and thermal feedback.
The locus of $\phi = 1 = \dfrac{k_o \rho_o \sqrt{g} A_o \sqrt{H_o}}{s \dot{m}''_{F,\infty} A_F}$

$$\therefore \text{at } \phi = 1, \ \frac{\rho_o \sqrt{g} A_o \sqrt{H_o}}{A_F} = \frac{s \dot{m}''_{F,\infty}}{k_o}$$

s and $\dot{m}''_{F,\infty}$ both are larger for liquids compared to wood.

CIB Data

In the CIB experiments, the fuel is placed over the entire floor; therefore, $A_F \approx A$.
Here, the theory gives

Fig. A.2 Wood crib and liquid pool fires

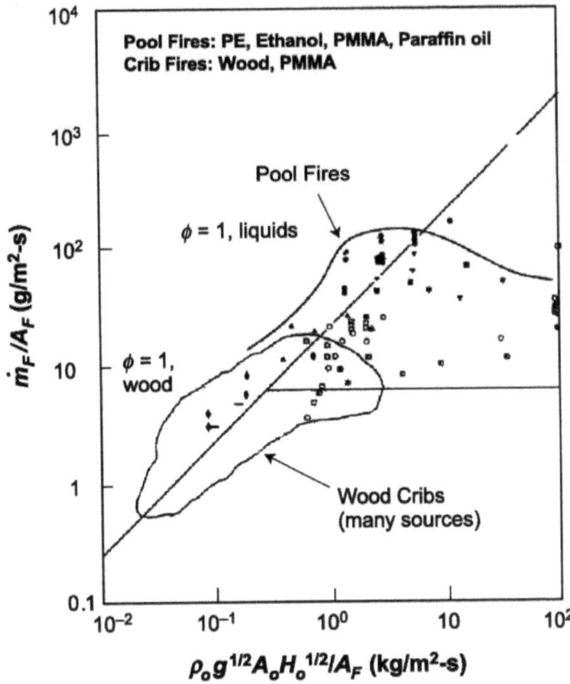

$$\phi > 1, \; \frac{\dot{m}_F}{A_o\sqrt{H_o}} = \frac{k_o\rho_o\sqrt{g}A_o\sqrt{H_o}}{sA_o\sqrt{H_o}} = \frac{k_o\rho_o\sqrt{g}}{s} = 0.125\,kg/_{sm^{5/2}} = 7.5\,kg/_{\min m^{5/2}}$$

From Eqs. 26a and 26b (Chapter "Fully Developed Enclosure Fires"), for $\phi < 1, T \approx \frac{A}{A_o\sqrt{H_o}}$ and for $\phi \geq 1, T \approx \frac{1}{\left(A/_{A_o\sqrt{H_o}}\right)}$ which agrees with the trends in Fig. A.2.

Eurocode

The Eurocode prescribes:

$$T = 1325\left(1 - 0.324e^{-0.2t*} - 0.204e^{-1.7t*} - 0.472e^{-19t*}\right)$$

$$t* = t\Gamma, \quad \Gamma = \left[\frac{\left(\frac{A_o\sqrt{H_o}}{A}\right)}{\sqrt{k\rho c}}\right]^2 \left(\frac{1160}{0.04}\right)^2$$

Here, $t*$ is essentially Q_w^{*-2} or $\frac{T-T_0}{T_0} = f(Q_w^*)$ only.

This specification must assume a ventilation-limited fire and ignores the other variables.

Lie

Only a ventilation-limited fire is assumed.

Lie gives $\dot{m}_F = 330 A_o \sqrt{H_o}$ (kg/h).

The theoretical development gives

$$\dot{m}_F = \frac{k_o \rho_o \sqrt{g} A_o \sqrt{H_o}}{s} = 450 A_o \sqrt{H_o}.$$

For temperature, $T = f(F, t)$ and C, a constant that takes into account the properties of the bounding materials of the enclosure.

$$F = \frac{A_o \sqrt{H_o}}{A}$$

This represents $k \rho c$ since $k \sim \rho$.

Magnusson, Thelandersson, and Petersson

Magnusson, Thelandersson, and Petersson compute a result for temperature based on a similar theory. They augment it with a rate of rise for developing fire and a prescribed cooling phase. They use only the ventilation-controlled fire for cribs from Kawagoe and Sekine:

$$\dot{m}_F = 0.092 A_o \sqrt{H_o}.$$

(The theoretical development gives $\dot{m}_F = 0.125 A_o \sqrt{H_o}.$)

They compute results for $\dfrac{A_o \sqrt{H_o}}{A}$ and $k \rho c$ for various fuel loads.

$\dfrac{\dot{Q}}{A} = \dot{Q}'' = \dot{m}_f'' \Delta H_c$ based on a distribution of fuel over the entire compartment surface area A. (Normally, fuel loading is based on floor area, i.e. $\dot{m}_F / A_{\text{floor}}$.)

Babrauskas

A computer solution was correlated to give an analytical result:

$$T - T_0 = 1432 K^* \theta_1^* \theta_2^* \theta_3^* \theta_4^* \theta_5$$
$$\theta_1 = 1 + 0.51 \ln\phi \text{ for } \phi < 1$$
$$\text{or } \theta_1 = 1.0 - 0.05 (\ln\phi)^{5/3} \text{ for } \phi > 1 \text{ for wood cribs}$$
$$\text{and } \theta_1 = 1.0 - 0.092 (-\ln\eta)^{1.25} \text{ for pool fires}$$

where:

$$\eta = \frac{A_o\sqrt{H_o}}{A_f} \frac{0.5\Delta H_c}{s\sigma(T^4 - T_b^4)}$$

Here, $\eta \sim Q_F^{*-1}\left(\frac{\Delta H_c}{L}\right)$

$$\theta_2 = 1.0 - 0.94 \exp\left[-54\left(\frac{A_o\sqrt{H_o}}{A}\right)^{2/3}\left(\frac{\delta}{k}\right)^{1/3}\right]$$

$$\theta_3 = 1.0 - 0.92 \exp\left[-150\left(\frac{A_o\sqrt{H_o}}{A}\right)^{3/5}\left(\frac{t}{k\rho c}\right)^{2/5}\right]$$

Both θ_2 and θ_3 correspond to Q_w^*, but not exactly, since the powers are different in θ_3 for each term. Since the dimensionalization of the equations must be consistent, it suggests that there is an inconsistency in θ_3.

$$\theta_4 = 1.0 - 0.205 H^{-0.3}$$

This corresponds to $Q_r^* \sim \dfrac{1}{\sqrt{H_o}}$.

θ_5 pertains to combustion efficiency and is only relevant if the theoretical heat of combustion is used.

It is interesting that the maximum temperature given by the correlation is 1425 °C. The theory suggests this is 1500 °C at most.

Law

Law developed a correlation based on the CIB data [16]. A fit giving the maximum or upper values of data is

$$T_{(\max)} = 6000 \left(\frac{1 - e^{-0.1\frac{A}{A_o\sqrt{H_o}}}}{\sqrt{\frac{A}{A_o\sqrt{H_o}}}} \right)$$

where A is the heat transfer area of the boundary surfaces, not including the vents (as used in the theory).

An adjustment is made if the fuel load is low.

The mass loss rate is correlated as

$$\dot{m}_f = 0.18 A_o\sqrt{H_o}\left(\frac{W}{D}\right)\left(1 - e^{-0.036\frac{A}{A_o\sqrt{H_o}}}\right)$$

for $\dfrac{\dot{m}_F}{A_o\sqrt{H_o}}\left(\dfrac{D}{W}\right)^{1/2} < 60$

where:

D = Compartment depth
W = Compartment width

Ma and Mäkeläinen

These authors develop a correlation based on the CIB and other data. Its novel feature is that it includes a prediction of temperature over time starting at the onset of the fully developed stage. They use Harmathy's result for the burning rate in the fuel-controlled regime, and his demarcation of the regime change to ventilation-limited:

$$\dot{m}''_{F,\infty} = 6.2\,\mathrm{g/m^2 s}, \quad \frac{\rho_o\sqrt{g}A_o\sqrt{H_o}}{A_F} \geq 0.263$$

They use Law's correlation for the ventilation-limited burning rate.

The temperature is given as

$$\frac{T - T_o}{T_m - T_o} = \frac{t}{t_m}\exp\left(1 - \frac{t}{t_m}\right)^{\delta}$$

where:

$\delta = 0.5$ for the ascending phase and
     1.0 for the descending phase

The maximum temperature is given as linear fits to tire CIB and other data in terms of $A/A_o\sqrt{H_o}$. The time at the maximum temperature is selected as $t_m = 0.63 m_f/\dot{m}_F$.

This model does not include the effect of the wall's thermal properties.

Appendix B: Comparisons of Enclosure Fire Predictions with Data

Predictions of compartment fire temperature and duration are compared to two sets of data. The first set of data is from 321 experiments conducted under the auspices of CIB [48]. See the section entitled CIB beginning on page 31 for more information on these experiments. The compartments in these experiments were roughly cubic, although some of the compartments had aspect ratios (length to width) of $1/2$ or 2.

In these experiments, the stage of fully developed burning was defined as the period from when the mass of fuel was between 80% and 30% of the original, unburned fuel mass. Average temperatures during the period of fully developed burning from these experiments were presented as a function of $\frac{A}{A_o\sqrt{H_o}}$.

Average burning rate data during the fully developed stage was presented as $\frac{\dot{m}_f}{A_o\sqrt{H_o}}\sqrt{\frac{D}{W}}$ as a function of $\frac{A}{A_o\sqrt{H_o}}$. Data was also included where the average burning rate during the fully developed burning stage was presented in tables of $\frac{\dot{m}_f}{A_o\sqrt{H_o}}$ as a function of $\frac{A}{A_o\sqrt{H_o}}$. Although both the CIB report [48] and the Cardington data [103] show that the aspect ratio of a compartment can influence the burning rate for fully developed, ventilation-limited fires, most predictive methods do not explicitly account for this effect. Therefore, predictive methods that do not account for compartment aspect ratio were evaluated using the CIB burning rate data, which was normalized by the area and square root of the height of the ventilation opening, but not by the square root of the ratio of compartment depth to width. Methods that do specifically account for the compartment aspect ratio were evaluated using the CIB data that was normalized by both the area and square root of the height of the ventilation opening and the square root of the ratio of compartment depth to width. When the CIB data was not normalized by the square root of the ratio of compartment depth to width, there was more scatter in the data.

The methods presented in this guide were evaluated by plotting predictions of average temperature during the fully developed stage along with the CIB data. When comparing predictions to data, averages were taken of what appeared to be the fully developed stage from the temperature data. Similarly, predictions of duration were compared to the CIB data by dividing the initial mass of fuel, m_f, by the predicted duration, τ, and plotting this quantity along with the CIB data.

Some of the predictive methods required as input the surface area of the fuel. The ratio of fuel surface area to total room surface area (defined as including the area of the ceiling and walls, but not the area of the ventilation opening or the floor) was calculated for each of the CIB experiments. The average ratio of fuel surface area to total room surface area in these experiments was 0.75, with a standard deviation of 0.90. Figure B.1 shows a histogram of the ratio of fuel surface area to the enclosure surface area for the CIB experiments. For methods that require as input the fuel surface area, the value of $0.75A$ was used for comparing predictions to the CIB data.

To explicitly analyze the effect of long, narrow compartments, temperature data as a function of time from a series of experiments that were conducted in a

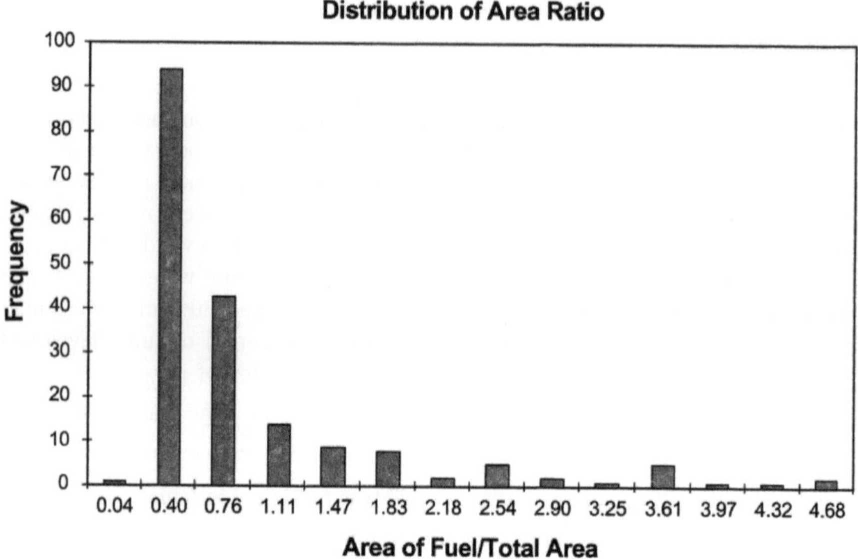

Fig. B.1 Histogram of ratio of fuel surface area to enclosure surface area for the CIB experiments

compartment that was approximately 23 m long, 2.7 m high, and 5.5 m wide [103] were compared to predictions. In these experiments, the ventilation opening ranged from 1/8 to 1/1 of the small side of the compartment. The fuel loading consisted of wood cribs with a total density of 20 or 40 kg/m². Additionally, for one experiment, the compartment size was reduced to approximately 5.6 × 5.6 × 2.75 m (high). The full details of the experiments may be found in reference 103.

CIB Data

The experiments in the CIB study were conducted in a variety of enclosures since multiple laboratories participated. Statistical means were used to overcome systematic differences between the laboratories. The majority of the laboratories used a test enclosure constructed of 10 mm thick asbestos millboard with a reported thermal conductivity of 0.15 W/m°C, and this is the value that was used for methods that required specific heat as an input. The density of the asbestos millboard and the specific heat were not reported, so values of 816 J/kg°C and 1100 kg/m³ were selected [27, 104].

In the CIB study, separate graphs of temperature and burning rate data were presented for cribs with 20 mm thick wood sticks spaced 20 mm apart, and for cribs with 20 mm wide sticks spaced 60 mm apart, or with 10 mm wide sticks spaced 30 mm apart. However, for purposes of comparing predictions with the CIB data, all temperature and burning rate data was aggregated into single graphs.

Cardington Data

A total of nine experiments were conducted under a collaborative project between British Steel and the British Research Establishment's Fire Research Station. The experiments were conducted in a purpose-built compartment within the British Research Establishment's ex-airship hanger.

The floor of the compartment was made of 75 mm thick concrete covered with sand. The walls were made of lightweight concrete blocks that measured $440 \times 215 \times 215$ mm. In most tests, the walls were lined with a 50 mm thick ceramic fiber blanket. However, in one of the tests (test #8) the walls were lined with two 12.5 mm thick plaster-board sheets affixed onto 47×47 mm wood studs spaced 600 mm apart. The ceiling was constructed of 200 mm thick aerated concrete slabs and was lined in the same manner as the walls.

The opening of the compartment was located on one of the smaller walls, and concrete blocks were used to restrict the opening to 100%, 50%, 25%, or 12.5% of the wall size. Additionally, in some of the tests, a 400 mm insulated steel column was placed flush with the opening, which further reduced the opening size.

The dimensions of the enclosure are provided in Table B.1 [103], the dimensions of the opening are listed in Table B.2 [103], and the properties of the enclosure materials are listed in Table B.3 [103].

Table B.1 Compartment dimensions of the Cardington tests

Test#	Length (m)	Width (m)	Height (m)
1	22.855	5.595	2.750
2	22.855	5.595	2.750
3	22.855	5.595	2.750
4	22.855	5.595	2.750
5	22.855	5.595	2.750
6	22.855	5.595	2.750
7	5.595	5.595	2.750
8	22.780	5.465	22.780
9	22.855	5.595	2.750

Table B.2 Opening dimensions of the Cardington tests

Test#	Total width (mm)	Height (mm)
1	5595	2750
2	5595	2750
3	5195	1470
4	5195	1470
5	2139	1730
6	5195	375
7	1370	2750
8	5065	2680
9	5195	2750

Table B.3 Properties of enclosure materials

Structure	Material	Density (kg/m^3)	Specific heat (J/kg K)	Thermal conductivity (W/m K)
Walls	Lightweight concrete blocks	1375	753	0.42
Roof	Aerated concrete slabs	450	1050	0.16
Floor	Sand	1750	800	1.0
Fiber lining	Ceramic fiber	128	1130	0.02
Plasterboard lining	Fireline plasterboard	900	1250	0.24

Table B.4 Fuel loading for the Cardington tests

Test #	Fuel load (kg/m^2)
1	40
2	20
3	20
4	40
5	20
6	20
7	20
8	20.6
9	20

The fuel for the Cardington tests was wood cribs, constructed of 1 m long sticks of 50×50 min western hemlock spaced 50 mm apart. The heat of combustion of the wood was reported as 19.0 MJ/kg. The fuel loading for each of the tests can be found in Table B.4.

In all, but tests #7 and #9, the fires were ignited at the rear of the compartment (opposite the end with the ventilation opening). In tests #7 and #9, all cribs were ignited simultaneously. In all the tests, the fire spread to the cribs nearest the ventilation opening, and, once the fire reached the cribs nearest the ventilation opening, the cribs further away from the ventilation opening ceased burning. The cribs nearest the ventilation opening continued burning, and, as the fuel was depleted, the fires progressed toward the rear of the enclosure. As a result, the temperatures were not horizontally homogeneous, and higher temperatures at any given time were measured above the location where the fire was burning.

The temperature data from the Cardington tests was compared to predictions made using the methods identified in this guide by comparing the measured temperatures to predictions. Temperatures were measured at locations approximately 3, 11, and 19 m (measured horizontally) from the ventilation opening. In the graphs, averages of the thermocouple measurements are plotted, with error bars indicating the range of the measured temperatures.

Predictions were made using each of the methods identified in this guide at 3-min intervals for tests #1, 2, 3, 7, and 9; at 6-min intervals for tests #4, 5, and 8; and at 25-min intervals for test #6.

For predictive methods that have distinct correlations for fuel-controlled and ventilation-controlled burning, the fire was assumed to be ventilation-controlled. Given the behavior of the burning, this is a reasonable assumption.

Eurocode

CIB Data

In the CIB experiments, the mass of fuel per unit area ranged from 20 to 40 kg/m^2. (A few tests used a mass of fuel per unit area of 10 kg/m^2, but, since the CIB report indicated that only a "few" tests were conducted at this density, this value was not modeled.) For an effective heat of combustion for pine of 12.4 MJ/kg [33], $q_{t,f}$ would range from 248 to 496 MJ/m^2, and multiplying this by the ratio of A_{floor}/A in the CIB compartments results in a range of $q_{t,d}$ of approximately 50 to 100 MJ/m^2.

Predictions of temperature as a function of time were made using the Eurocode method for values of $\frac{A}{A_o\sqrt{H_o}}$ ranging from 5 to 50 m$^{-1/2}$. Predictions were made at time increments ranging from 0.005 to 5 h, depending on the values of $q_{t,\,d}$ and $\frac{A}{A_o\sqrt{H_o}}$ used. For each value of $\frac{A}{A_o\sqrt{H_o}}$, averages of the temperature predictions during the time in which $t^* < t_d^*$ were compared to the CIB data. The Eurocode method was evaluated as presented, and the modifications suggested by Buchannan and Franssen also were evaluated. A graph of Eurocode predictions and the CIB data is presented in Figs. B.2 and B.3.

The predicted duration of the fully developed burning stage is when $t^* = t_d^*$. Given that $t_d^* = \dfrac{0.13 \times 10^{-3} q_{t,d} \Gamma A}{A_o\sqrt{H_o}}$, and $t^* = t\Gamma$, the predicted duration in hours would be $\tau = \left[0.13 \times 10^{-3} q_{t,d} \Gamma / \left(\frac{A_o\sqrt{H_o}}{A} \right) \right] \div \Gamma$, where τ is in hours and can be rewritten as $\tau = 0.13 \times 10^{-3} q_{t,d} \frac{A}{A_o\sqrt{H_o}}$.

Substituting $q_{t,d} = q_{f,d} \frac{A_{\text{floor}}}{A}$,

$$\tau = 0.13 \times 10^{-3} q_{f,d} \frac{A_{\text{floor}}}{A} \frac{A}{A_o\sqrt{H_o}}$$

Since $\dot{m}_f = \frac{m_f}{\tau}$,

$$\dot{m}_f = \frac{m_f}{0.13 \times 10^{-3} q_{f,d} \frac{A_{\text{floor}}}{A} \frac{A}{A_o\sqrt{H_o}}}.$$

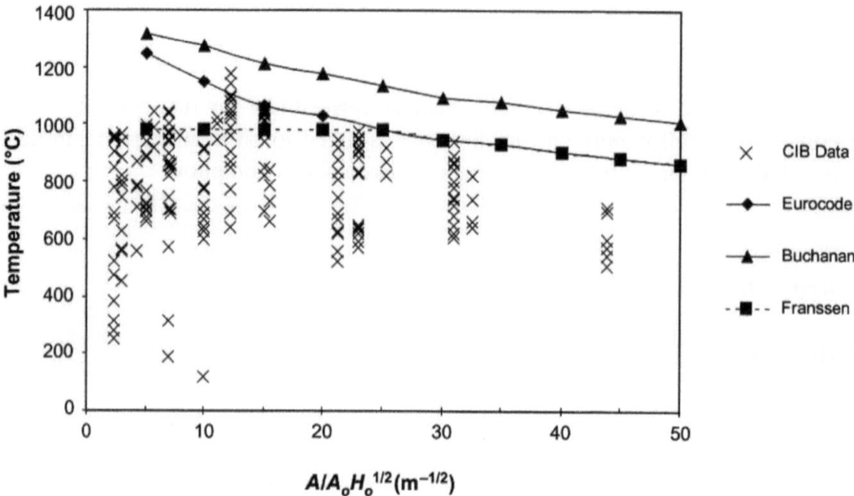

Fig. B.2 Comparison of CIB temperature data to predictions made using Eurocode, Buchanan, and Franssen methods, $q_{t,d} = 100$ MJ/m^2

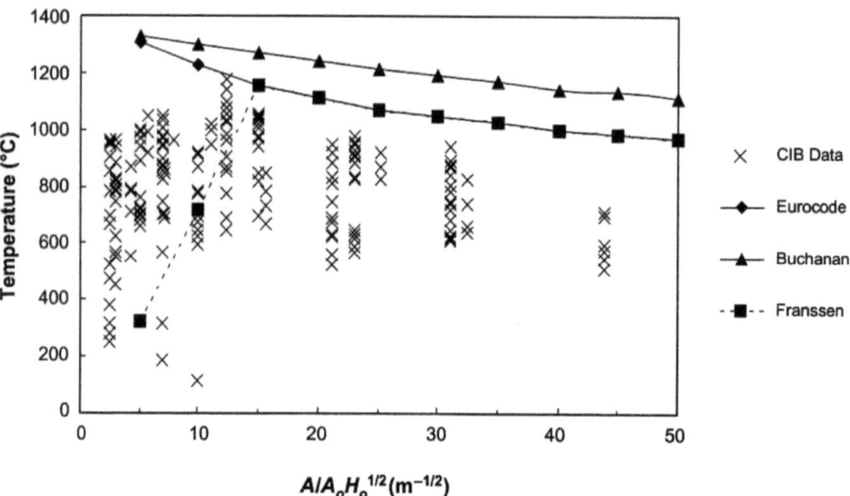

Fig. B.3 Comparison of CIB temperature data to predictions made using Eurocode, Buchanan, and Franssen methods, $q_{t,d} = 50$ MJ/m^2

Since $q_{f,d} = m''_f \Delta H_c$ and $m''_f = \dfrac{m_f}{A_{\text{floor}}}$, $\dot{m}_f = \dfrac{7700 m_f}{\frac{m_f \Delta H_c}{A_{\text{floor}}} \frac{A_{\text{floor}}}{A} \frac{A}{A_o \sqrt{H_o}}}$, which can be rearranged as $\dot{m}_f = 7700 \dfrac{A_o \sqrt{H_o}}{\Delta H_c}$ (kg/h) or $\dot{m}_f = 2.1 \dfrac{A_o \sqrt{H_o}}{\Delta H_c}$ (kg/s). Substituting $\Delta H_C = 12.4$ MJ/kg, the predicted burning rate would be $\dot{m}_f = 0.17 A_o \sqrt{H_o}$. This is compared to the CIB burning rate data in Fig. B.4.

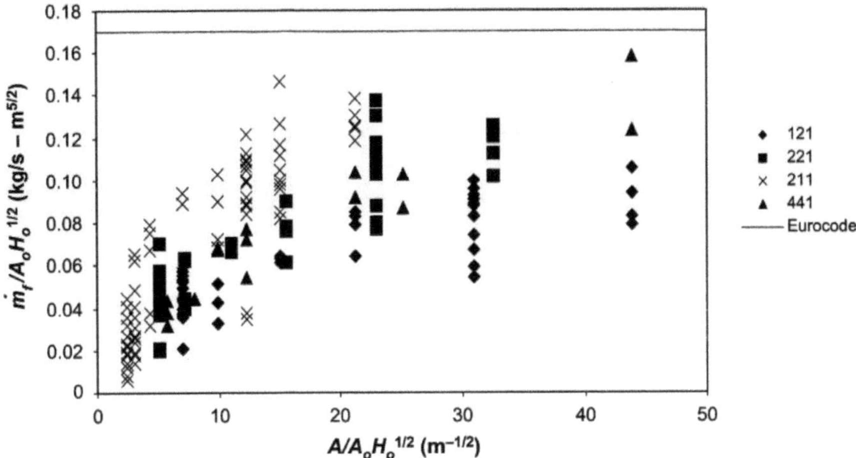

Fig. B.4 Comparison of CIB burning rate data to predictions made using Eurocode method

Franssen's modification results in a calculated burning duration of 20 min when t^*_d/Γ is less than 20 min. For the CIB data and $q_{t,d} = 50$ MJ/m^2, t^*_d/Γ is less than 20 min for cases where $\frac{A}{A_o\sqrt{H_o}}$ was less than or equal to 10 m$^{-1/2}$. With $q_{t,d} = 100$ MJ/m^2, t^*_d/Γ is less than 20 min for cases where $\frac{A}{A_o\sqrt{H_o}}$ was less than or equal to 30 m$^{-1/2}$.

Cardington Data

Inputs were created in accordance with the recommendations of the Eurocode. When calculating $q_{t,d}$, the area of the ventilation opening was not included in the calculation of the total surface area; however, the area of the openings was included in calculations of the total surface area of the enclosure. Predictions less than 20 °C were assumed to indicate that the decay period had been completed, and the temperature in the compartment was ambient. The results of the comparisons of predictions using the Eurocode to the Cardington data are presented in Figs. B.5, B.6, B.7, B.8, B.9, B.10, B.11, B.12 and B.13.

Lie

Since it was not possible to determine the duration of burning for each data point in the CIB data in a straightforward manner, to compare predictions using Lie's method to the CIB data average temperature predictions were made for a fire of 2 h' duration

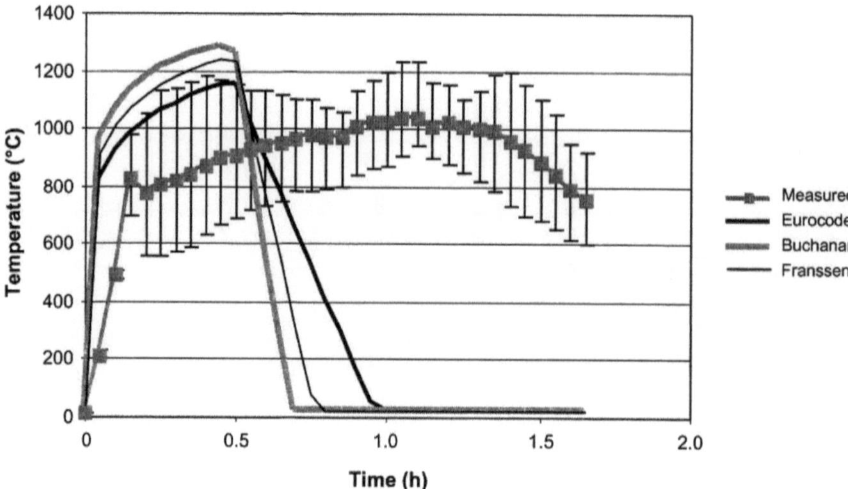

Fig. B.5 Comparison of predictions made using Eurocode, Buchanan, and Franssen methods to data from Cardington test #1

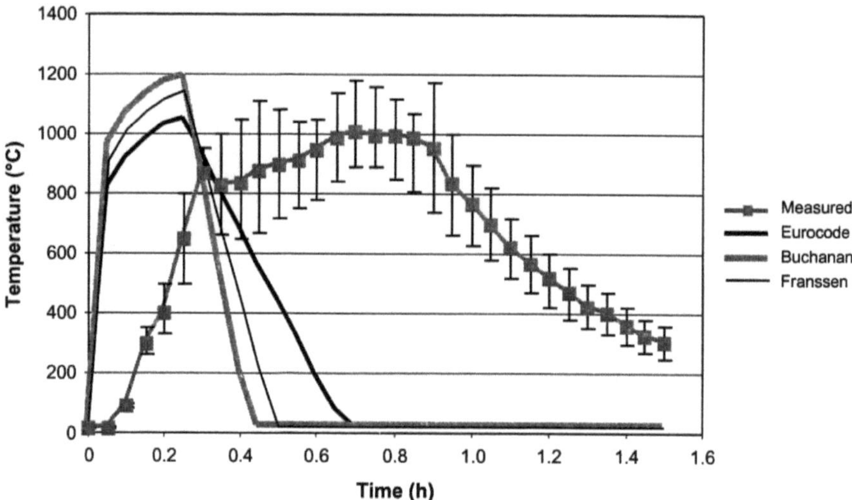

Fig. B.6 Comparison of predictions made using Eurocode, Buchanan, and Franssen methods to data from Cardington test #2

with opening factors $F = \frac{A_o \sqrt{H_o}}{A}$ ranging from 0.02 to 1. Because the density of the enclosures used in the CIB tests was assumed to be 1100 kg/m^3, the C factor used in Lie's method would equal 1. A comparison of Lie's predictions and the CIB data can be found in Fig. B.14.

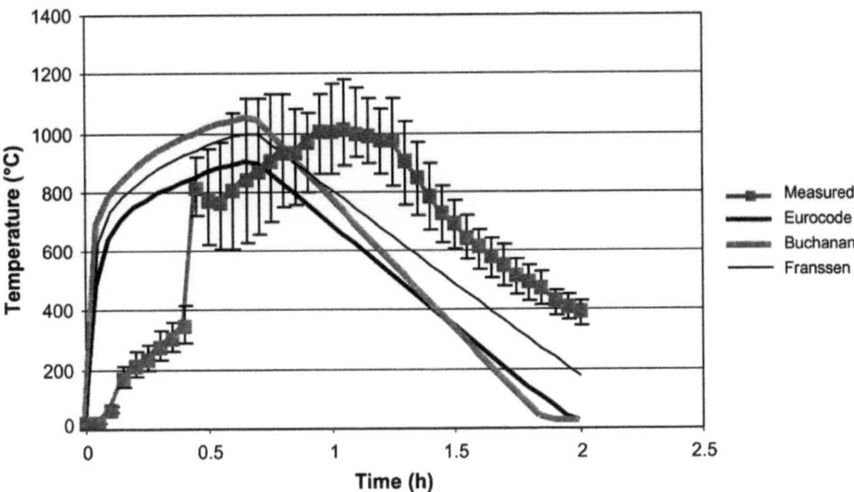

Fig. B.7 Comparison of predictions made using Eurocode, Buchanan, and Franssen methods to data from Cardington test #3

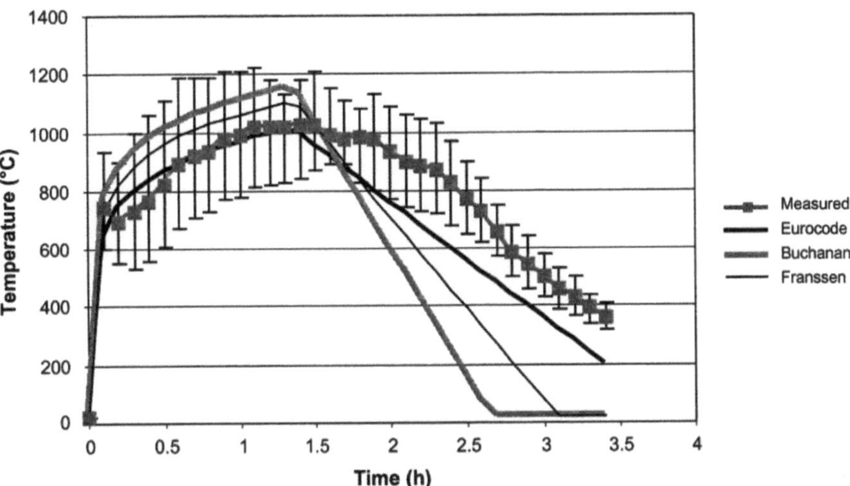

Fig. B.8 Comparison of predictions made using Eurocode, Buchanan, and Franssen methods to data from Cardington test #4

Lie gives $\dot{m}_f = 0.092 A_o \sqrt{H_o} (\mathrm{kg/s})$. This is compared to the CIB burning rate data in Fig. B.15.

Comparisons of predictions using Lie's method to the Cardington data can be found in Figs. B.16, B.17, B.18, B.19, B.20, B.21, B.22, B.23 and B.24.

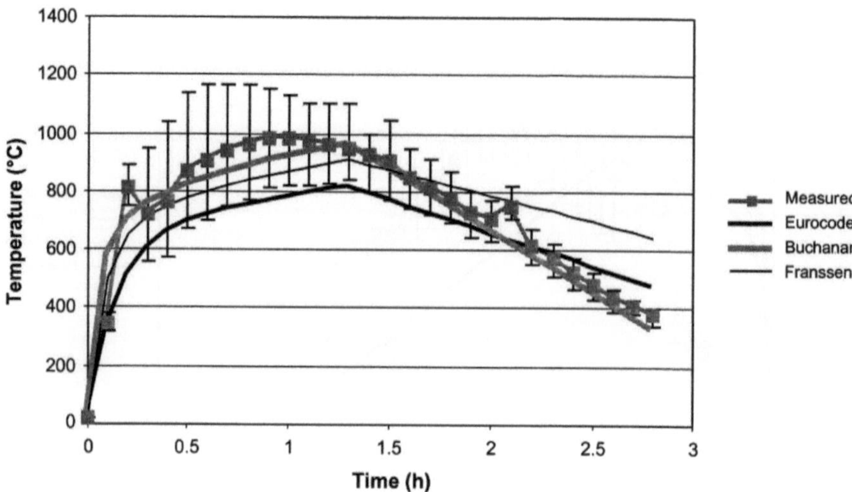

Fig. B.9 Comparison of predictions made using Eurocode, Buchanan, and Franssen methods to data from Cardington test #5

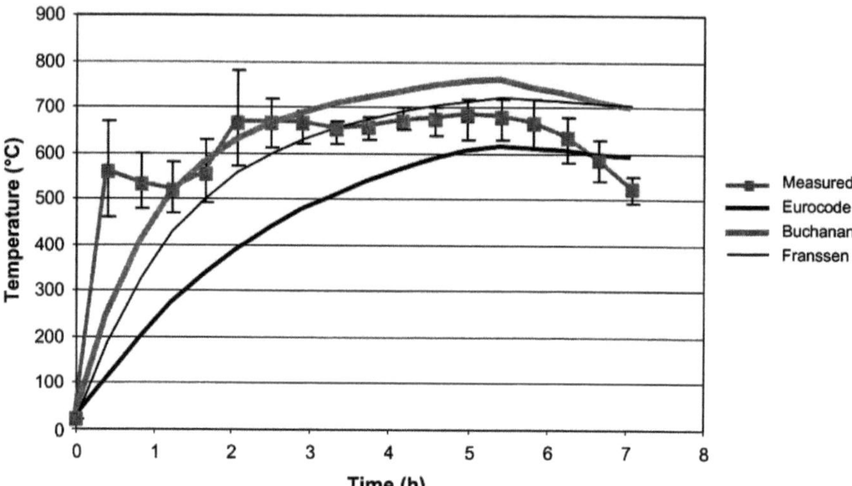

Fig. B.10 Comparison of predictions made Using Eurocode, Buchanan, and Franssen methods to data from Cardington test #6

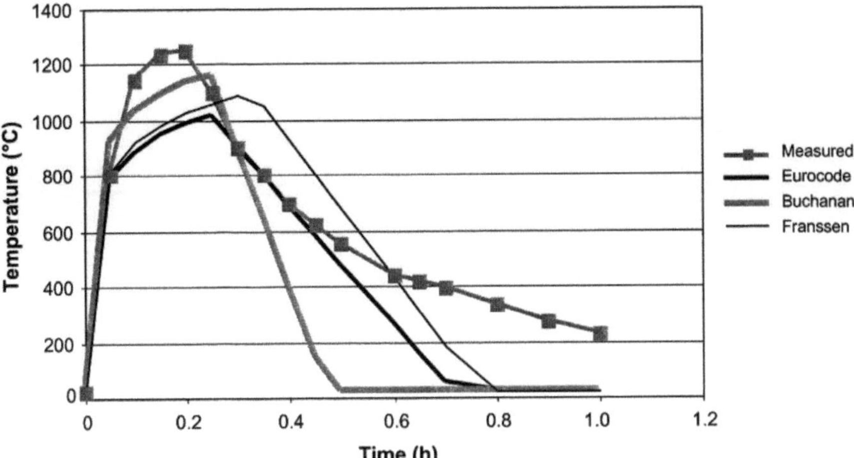

Fig. B.11 Comparison of predictions made using Eurocode, Buchanan, and Franssen methods to data from Cardington test #7

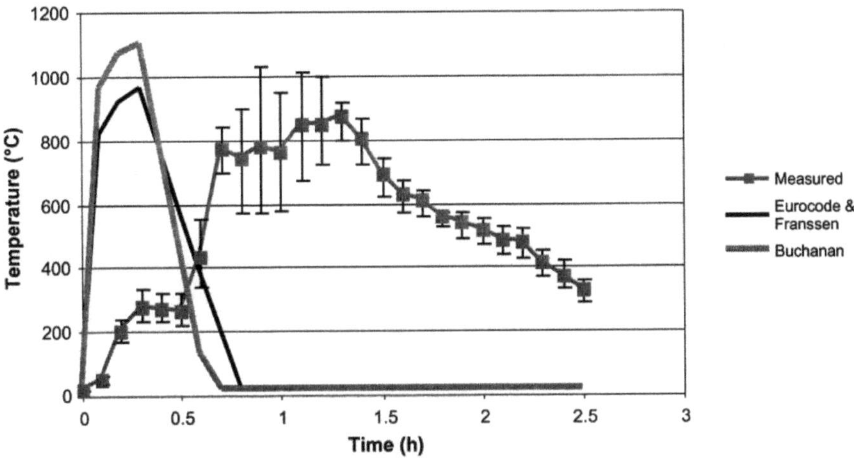

Fig. B.12 Comparison of predictions made using Eurocode, Buchanan, and Franssen methods to data from Cardington test #8

Tanaka

For Tanaka's methods, it was not possible to determine the duration of burning for each point in the CIB data in a straightforward manner. To compare predictions using Tanaka's method and his refined method to the CIB data, average temperature predictions were made for a fire of 2 h' duration with $\frac{A}{A_o\sqrt{H_o}}$ ranging from 1 to 50 m$^{-1/2}$. For $\frac{A}{A_o\sqrt{H_o}} = 1$ m$^{-1/2}$, Tanaka's refined method produced rapidly declining

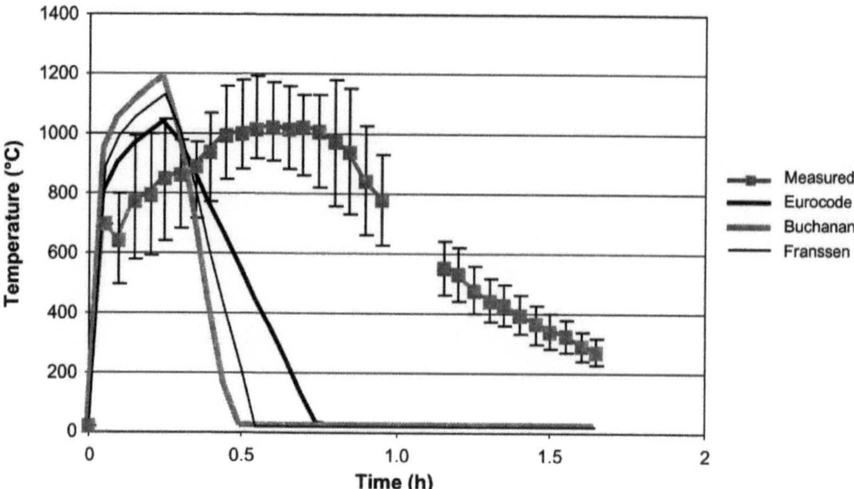

Fig. B.13 Comparison of predictions made using Eurocode, Buchanan, and Franssen methods to data from Cardington test #9

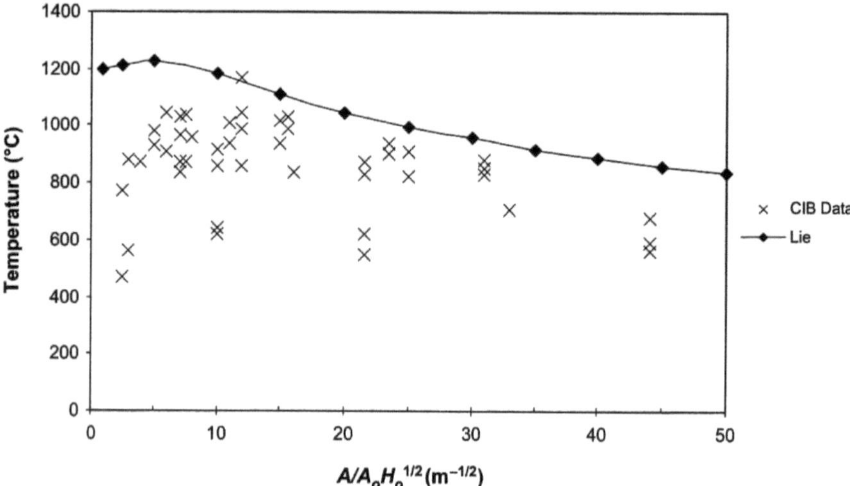

Fig. B.14 Comparison of CIB temperature data to predictions made using Lie's method

temperatures, and any temperature below 600 °C was neglected. The result of this comparison can be seen in Fig. B.25.

Both Tanaka's method and Tanaka's refined method predict the mass loss rate as $\dot{m}_f = 0.1 A_o \sqrt{H_o}$. This is compared with the CIB data in Fig. B.26.

Comparisons of predictions using Tanaka's method, both the simple and refined versions, to the Cardington data can be found in Figs. B.27, B.28, B.29, B.30, B.31, B.32, B.33, B.34 and B.35.

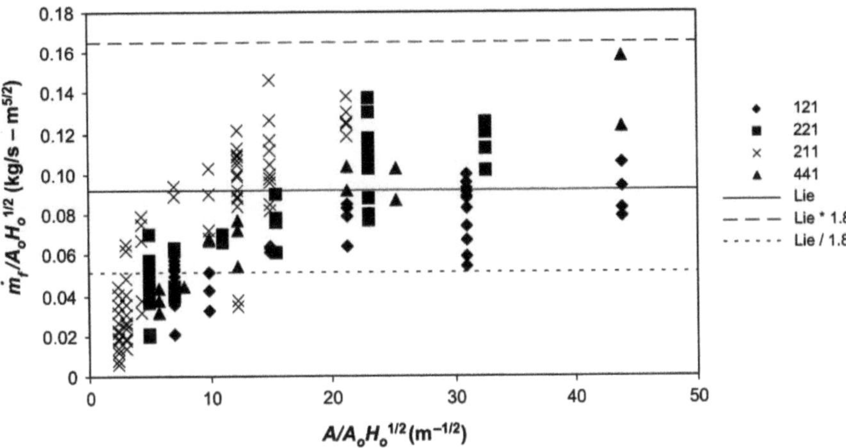

Fig. B.15 Comparison of CIB burning rate data to predictions made using Lie's method

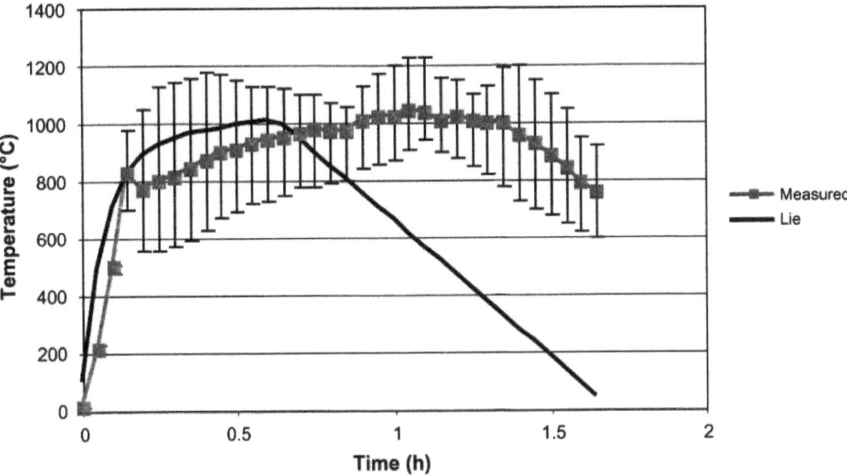

Fig. B.16 Comparison of predictions made using Lie's method to data from Cardington test #1

Magnusson and Thelandersson

The enclosures that were used in the CIB tests were modeled as Type C (as defined by Magnusson and Thelandersson [38]) since the Type C enclosure most closely represents the material properties of the CIB enclosures.

Given that it was not possible to estimate tire burning rates applicable to the CIB data in a straightforward manner, a duration of 2 h was arbitrarily selected. This selection should have only a minor influence on the comparison with the CIB data since only the average temperature during the fully developed stage is of interest. A

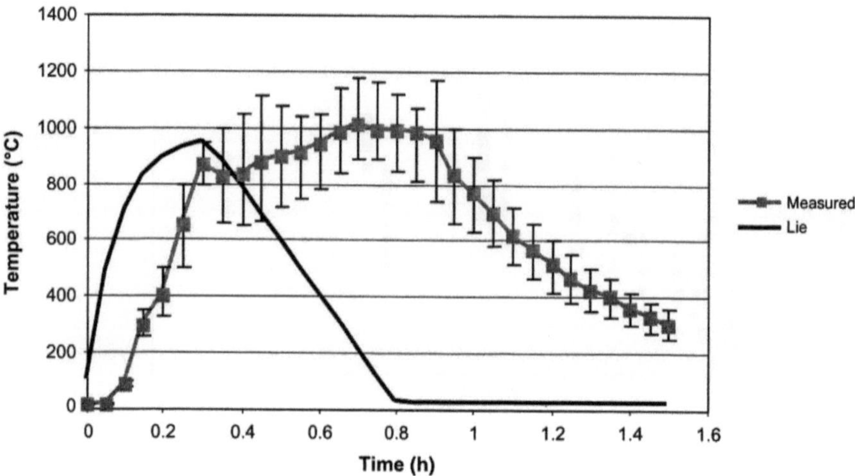

Fig. B.17 Comparison of predictions made using Lie's method to data from Cardington test #2

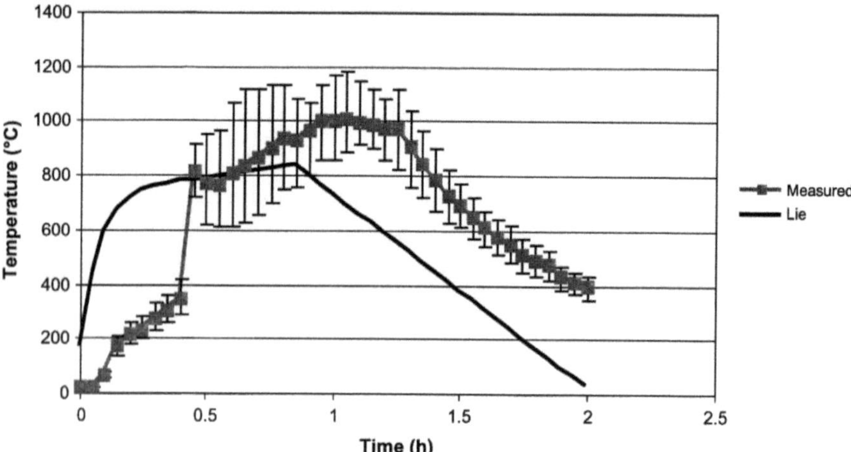

Fig. B.18 Comparison of predictions made using Lie's method to data from Cardington test #3

comparison of predictions made in this manner with the CIB data is shown in Fig. B.36.

Magnusson and Thelandersson's method predicts burning duration as follows:

$$\tau = \frac{qA}{25A_o \sqrt{H_o}} \; (\text{min})$$

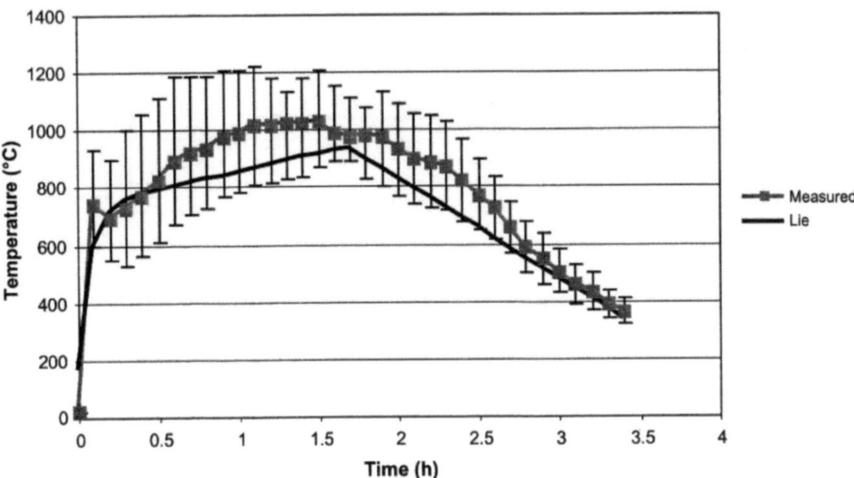

Fig. B.19 Comparison of predictions made using Lie's method to data from Cardington test #4

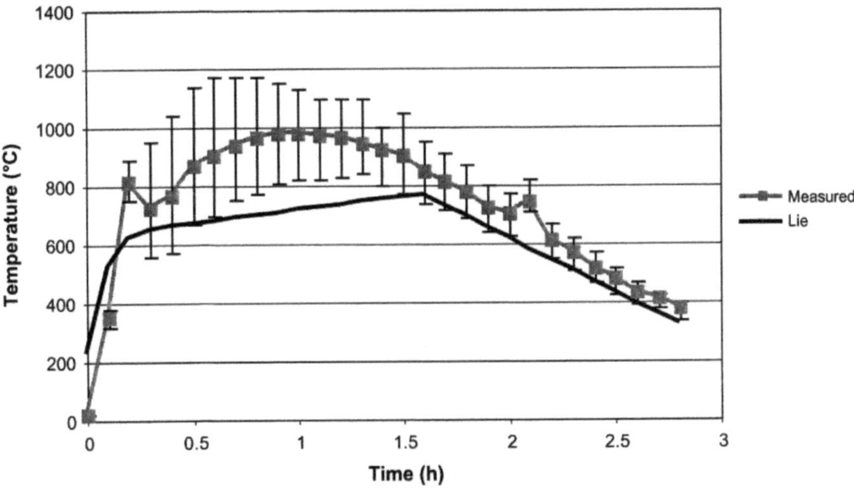

Fig. B.20 Comparison of predictions made using Lie's method to data from Cardington test #5

where q is the fuel load in Mcal/m^2 related to the surface area of the enclosure. Using a heat of combustion of 12.4 MJ/kg and converting units, this can be reduced to $\tau = 8.1 \frac{m_f}{A_o \sqrt{H_o}}$ (min).

Since $\dot{m}_f = \frac{m_f}{\tau}$, the burning rate predicted using Magnusson and Thelandersson's method would be $\dot{m}_f = 0.12 A_o \sqrt{H_o}$, which is identical to the method that Babrauskas recommends for ventilation-controlled burning. A comparison of predictions of burning rate made using Magnusson and Thelandersson's method to the CIB data is shown in Fig. B.37.

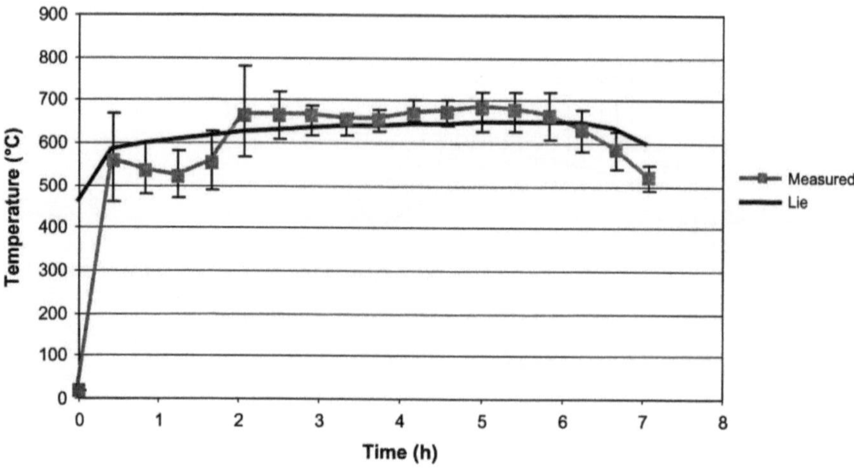

Fig. B.21 Comparison of predictions made using Lie's method to data from Cardington test #6

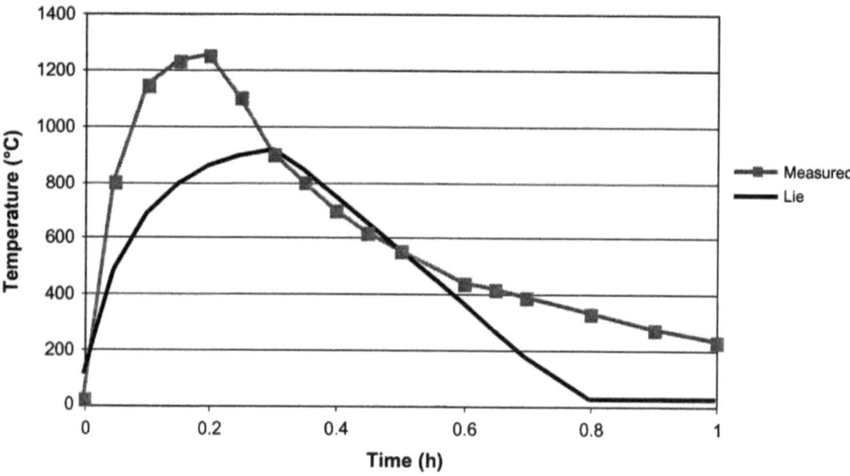

Fig. B.22 Comparison of predictions made using Lie's method to data from Cardington test #7

With the exception of test #8, which was modeled as Type G, the Cardington enclosure was modeled as Type C. The area of the ventilation opening was not included in calculations of the surface area of the enclosure. Where values of $\frac{A}{A_o\sqrt{H_o}}$ or the burning duration were not sufficiently close to the values presented in the tables, linear interpolation was performed. It was not possible to model test #6 using Magnusson and Thelandersson's method since no table or graph was provided that resembled the conditions associated with test #6. Comparisons of predictions made using Magnusson and Thelandersson's method to the Cardington data can be found in Figs. B.38, B.39, B.40, B.41, B.42, B.43, B.44 and B.45.

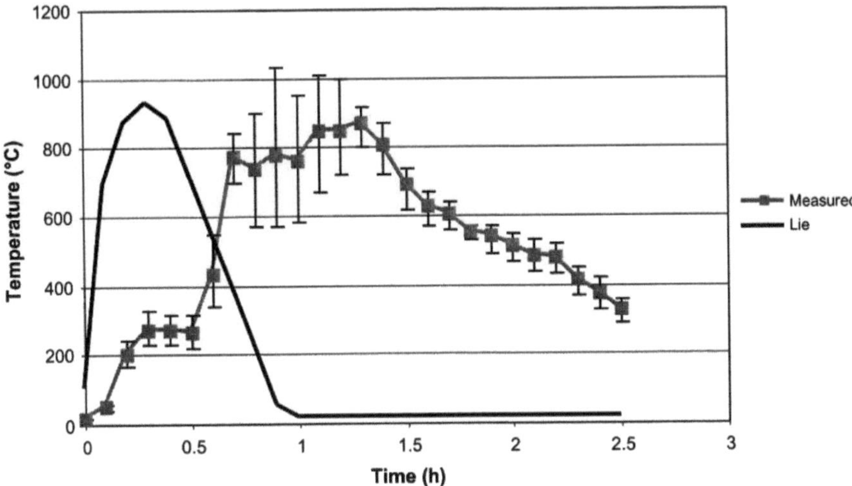

Fig. B.23 Comparison of predictions made using Lie's method to data from Cardington test #8

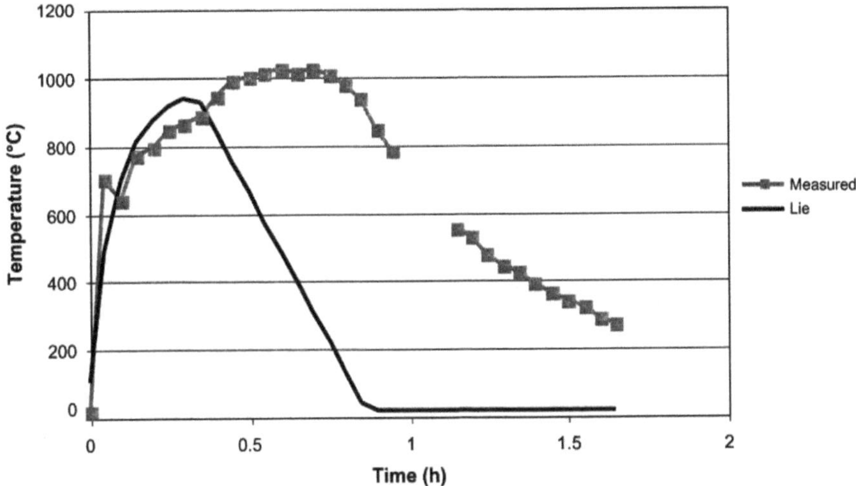

Fig. B.24 Comparison of predictions made using Lie's method to data from Cardington test #9

Harmathy

Because of the iterative nature of Harmathy's method for predicting compartment fire temperatures, it is not possible to compare predictions using Harmathy's method to the CIB data in a straightforward manner.

Harmathy distinguishes fuel-limited burning from ventilation-limited burning as the point where $\dfrac{\rho_0\sqrt{g}A_o\sqrt{H_o}}{A_f} = 0.263$. Substituting ρ_0 1.2 kg/m^3 and $g = 9.8$ m/s^2,

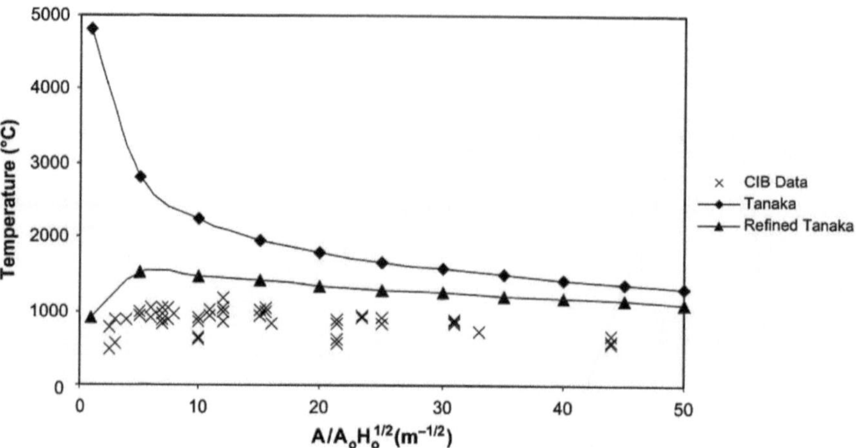

Fig. B.25 Comparison of CIB temperature data to predictions made using Tanaka's methods

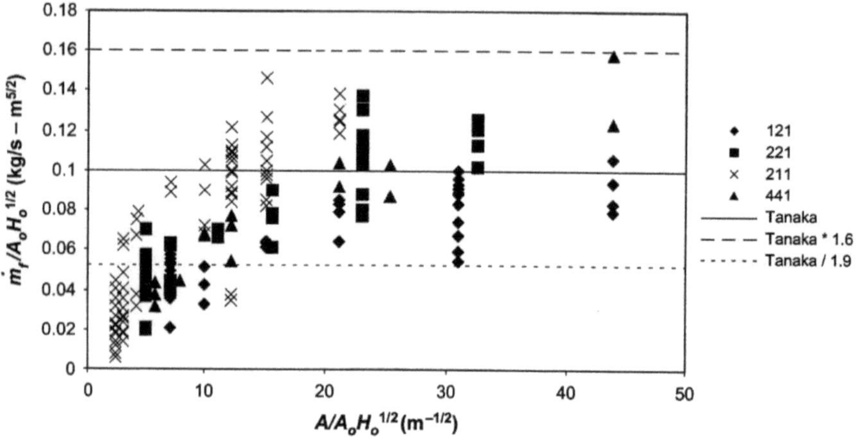

Fig. B.26 Comparison of CIB burning rate data to predictions made using Tanaka's methods

$\dfrac{A_o\sqrt{H_o}}{A_f} = 0.07$. In the CIB tests, the average value of AF/A was approximately 0.75. Substituting and inverting, the threshold between fuel-limited and ventilation-limited burning would be $\dfrac{A}{A_o\sqrt{H_o}} = 19.0$.

For fuel-limited burning, Harmathy gives: $\tau = \dfrac{151 m_f}{A_f}$. Substituting $A_f = 0.75A$ and $\dot{m}_f = m_f/\tau$ yields $\dot{m}_f = 0.00465A$.

For ventilation-limited burning, Harmathy gives: $\tau = 39.7 \dfrac{m_f}{\rho_0\sqrt{g}A_o\sqrt{H_o}}$.

Substituting $\rho_0 = 1.2$ kg/m^3 and $g = 9.8$ m/s^2, $\tau = 10.6 \dfrac{m_f}{A_o\sqrt{H_o}}$. Substituting this into $\dot{m}_f = m_f/\tau$ yields $\dot{m}_f = 0.09 A_o\sqrt{H_o}$. This is compared to the CIB data in Fig. B.46.

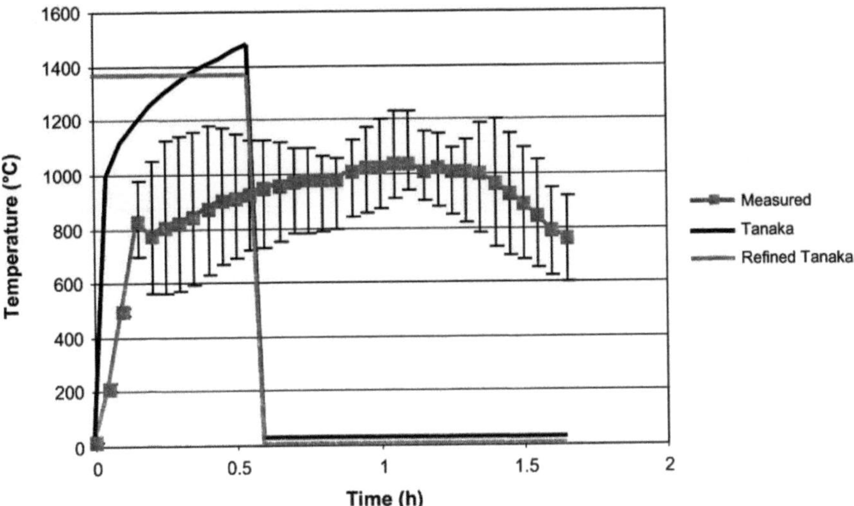

Fig. B.27 Comparison of predictions made using Tanaka's methods to data from Cardington test #1

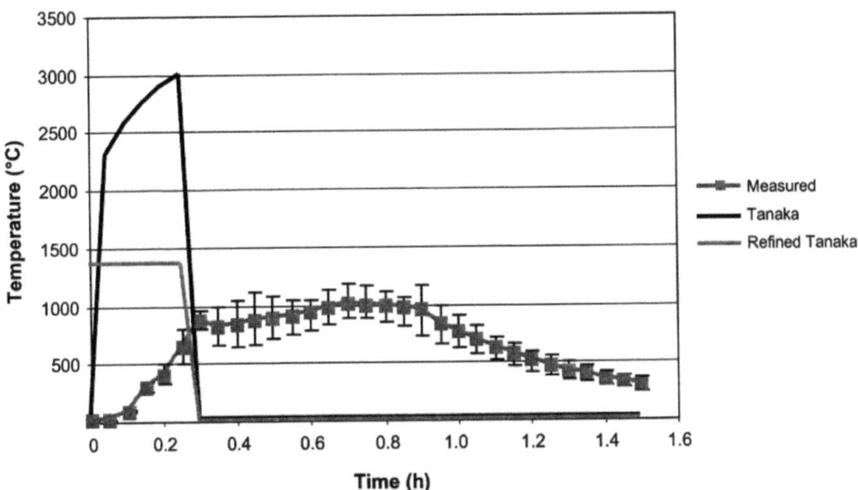

Fig. B.28 Comparison of predictions made using Tanaka's methods to data from Cardington test #2

Comparisons of predictions using Harmathy's method to the Cardington data are presented in Figs. B.47, B.48, B.49, B.50, B.51, B.52, B.53, B.54 and B.55. Predictions for times less than the burning duration were created by using the iterative method recommend by Harmathy, and a minimum resolution of 1 °C was required for the prediction to be accepted.

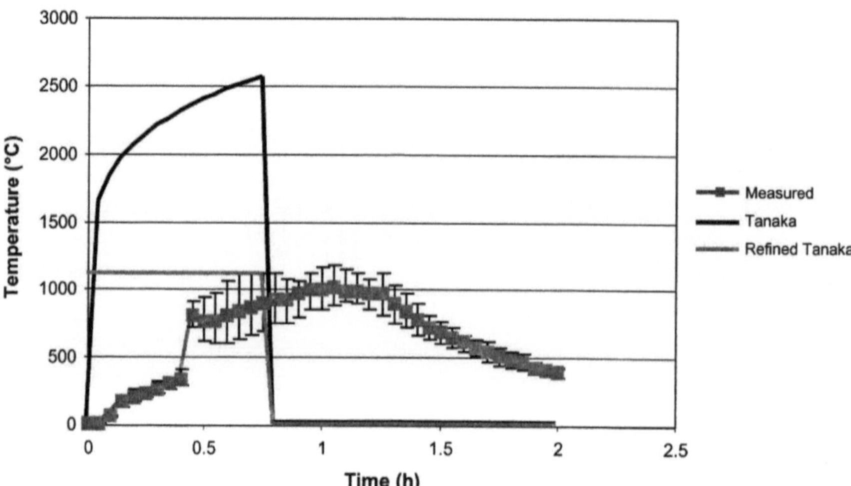

Fig. B.29 Comparison of predictions made using Tanaka's methods to data from Cardington test #3

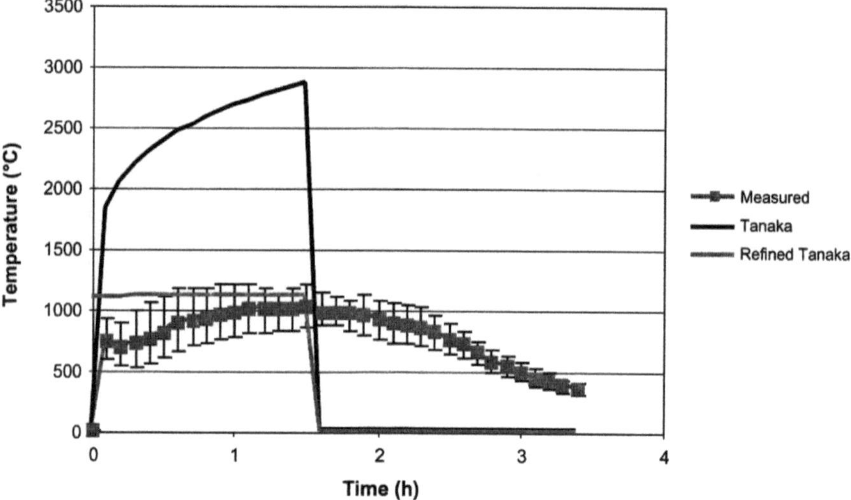

Fig. B.30 Comparison of predictions made using Tanaka's methods to data from Cardington test #4

Babrauskas

Babrauskas provides the equivalence ratio as $\phi = \dfrac{\dot{m}_f}{\dot{m}_{st}}$ where $\dot{m}_{st} = \dfrac{A_o\sqrt{H_o}}{2s}$ and s is the ratio such that 1 kg fuel + s kg air = $(1 + s)$ kg products. Harmathy [39] notes that a typical wood would have the chemical formula $CH_{1.455}O_{0.645} \cdot 0.233H_2O$, which

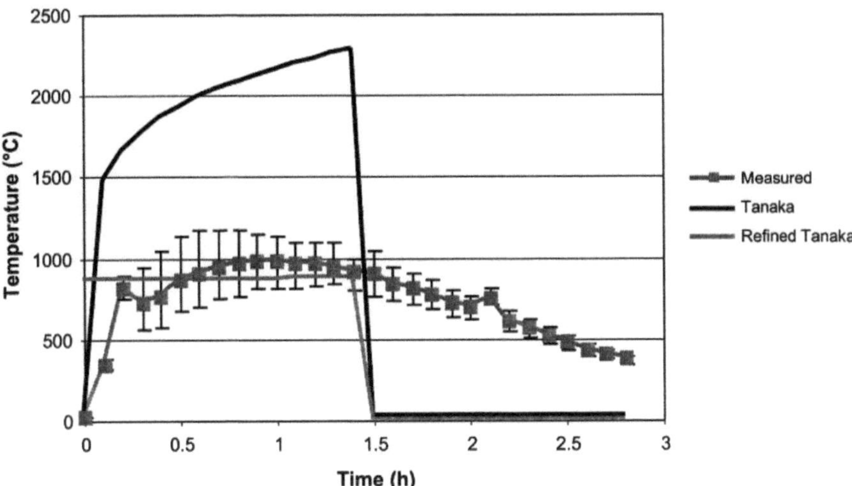

Fig. B.31 Comparison of predictions made using Tanaka's methods to data from Cardington test #5

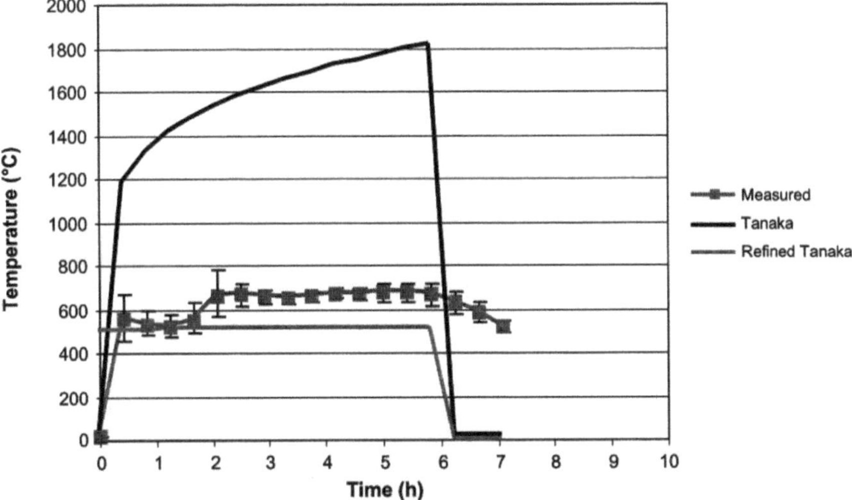

Fig. B.32 Comparison of predictions made using Tanaka's methods to data from Cardington test #6

would result in a value of s of 6.0, which is slightly larger than the value of 5.7 proposed by Babrauskas [46]. Using $s = 6.0$, $\dot{m}_{st} = 0.083 A_o \sqrt{H_o}$. Substituting this into the correlation for the equivalence ratio yields $\phi = \dfrac{\dot{m}_f}{0.083 A_o \sqrt{H_o}}$.

Babrauskas provides methods for modeling the burning rate for ventilation-controlled burning and for fuel-controlled burning, wood cribs, and thermoplastic

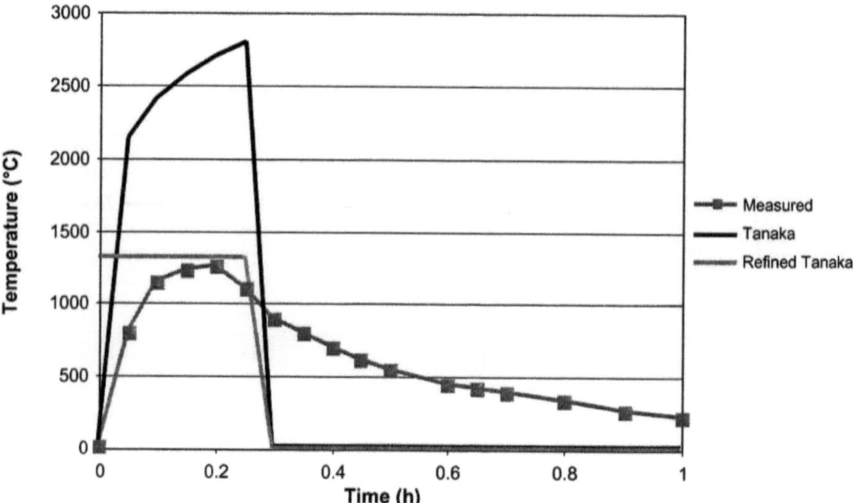

Fig. B.33 Comparison of predictions made using Tanaka's methods to data from Cardington test #7

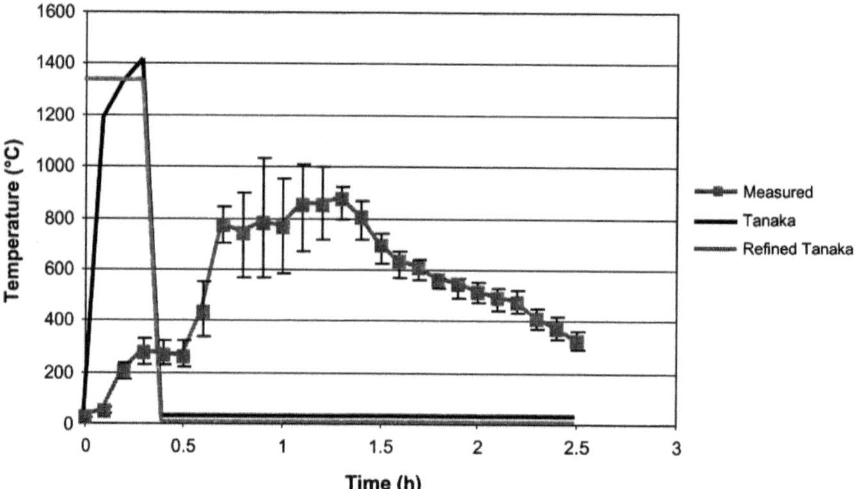

Fig. B.34 Comparison of predictions made using Tanaka's methods to data from Cardington test #8

or liquid pools [45]. Babrauskas' model for calculating the burning rate of ventilation-controlled fires is used here; however, in most design situations, the input data needed to use Babrauskas' models for fuel-controlled burning is not available. Therefore, Harmathy's model for the burning rate of over-ventilated fires was used for the present analysis.

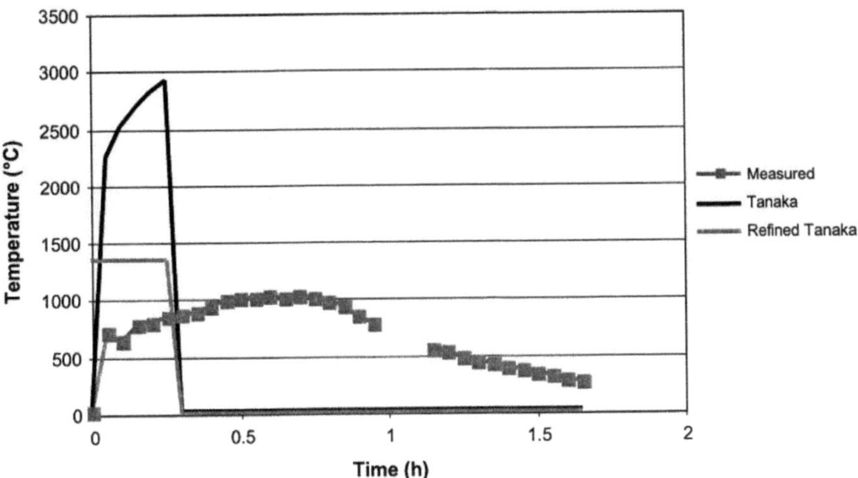

Fig. B.35 Comparison of predictions made using Tanaka's methods to data from Cardington test #9

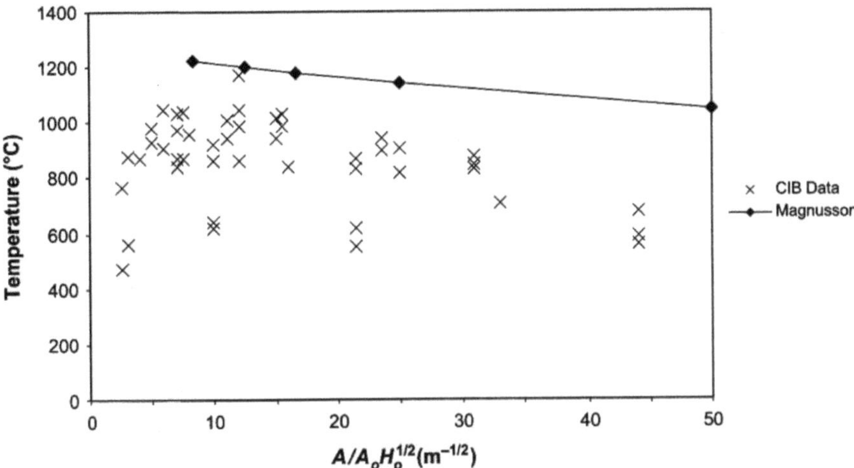

Fig. B.36 Comparison of CIB temperature data to predictions made using Magnusson and Thelandersson's method

For fuel-controlled burning, Harmathy estimates the burning rate as $\dot{m}_f = 0.0062A_f$. Substituting this into the above yields:

$\phi = 0.074 \dfrac{A_f}{A_o\sqrt{H_o}}$. For stoichiometric binning, $\phi = 1$. In the CIB tests, the average value of A_F/A was approximately 0.75. Substituting and solving for $\dfrac{A_o}{A_o\sqrt{H_o}}$, the

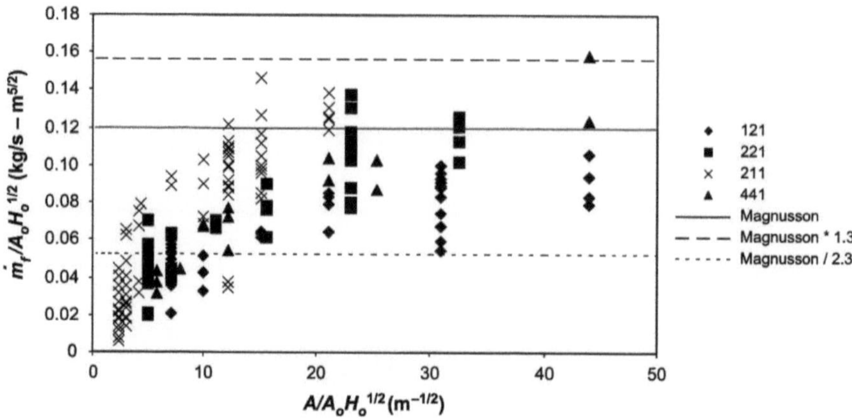

Fig. B.37 Comparison of CIB burning rate data to predictions made using Magnusson and Thelandersson's method

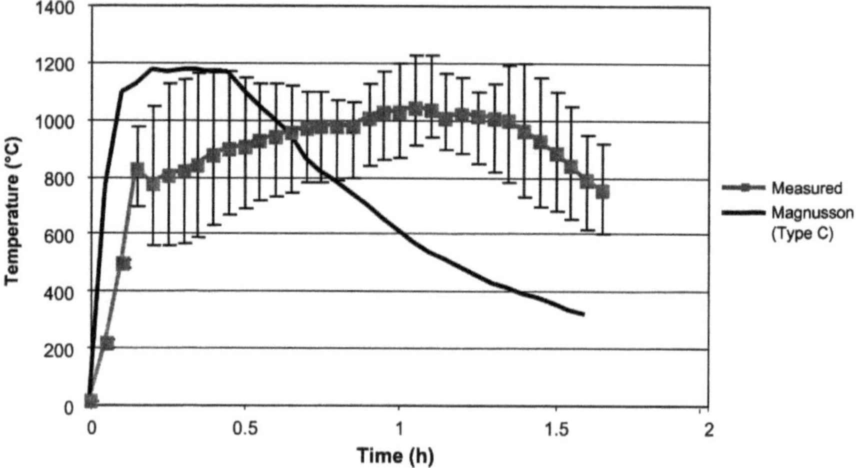

Fig. B.38 Comparison of predictions made using Magnusson and Thelandersson's method (Type C) to data from Cardington test #1

threshold between fuel-limited and ventilation-limited burning would be $\frac{A_o}{A_o\sqrt{H_o}} = 18.0$.

Substituting in the relevant values for enclosure properties from the CIB tests and assuming that $H_0 \approx 1$ m (in the CIB tests, H_o ranged from 0.5 to 1.5 m, but, given that Babrauskas' method varies with $H_o^{-0.3}$, predictions are not highly sensitive to this parameter) and $b_p = 0.9$ results in the predictions of the CIB temperatures shown in Fig. B.56.

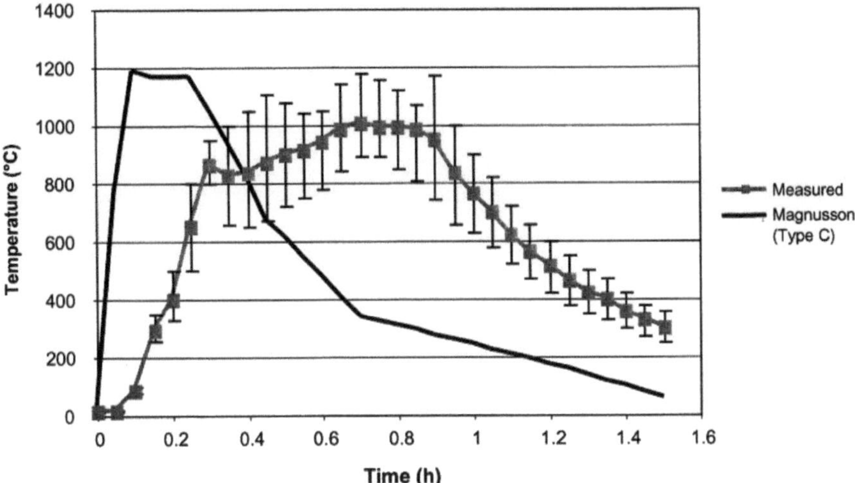

Fig. B.39 Comparison of predictions made using Magnusson and Thelandersson's method (Type C) to data from Cardington test #2

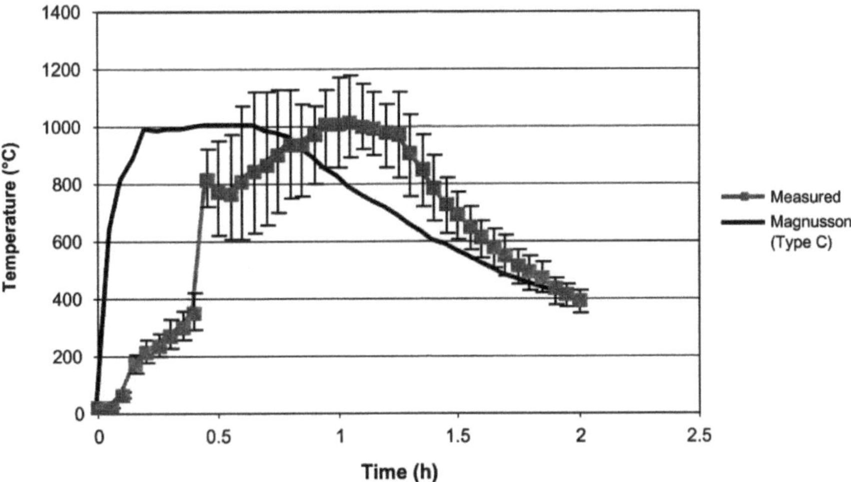

Fig. B.40 Comparison of predictions made using Magnusson and Thelandersson's method (Type C) to data from Cardington test #3

For ventilation-controlled burning, Babrauskas estimates the burning rate as [45]:

$$\dot{m}_f = 0.12A_o\sqrt{H_o}$$

Given that Harmathy's method of estimating burning rate for fuel-controlled burning was used, the evaluation of that method is applicable to the assumption

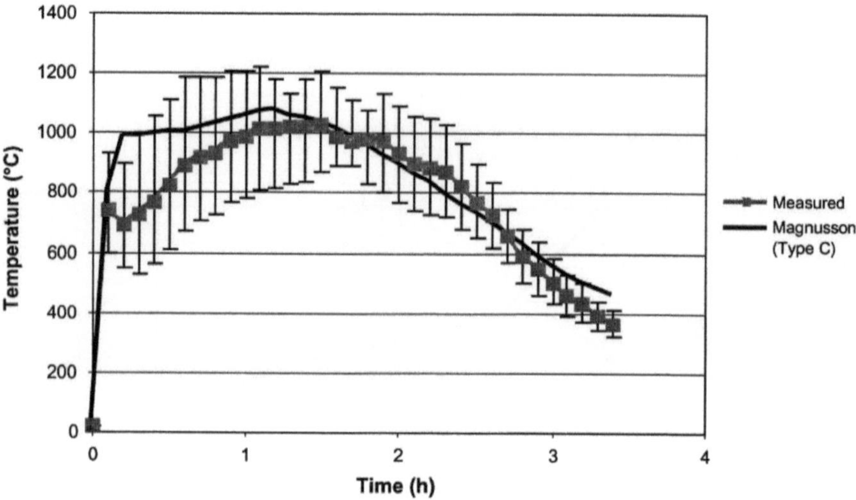

Fig. B.41 Comparison of predictions made using Magnusson and Thelandersson's method (Type C) to data from Cardington test #4

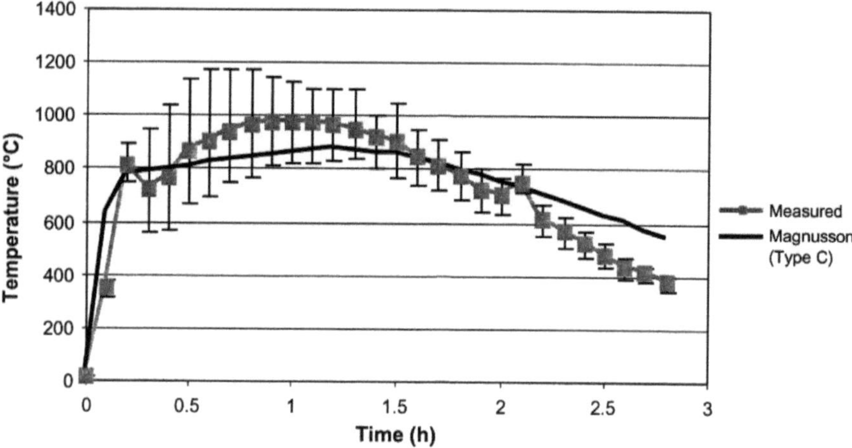

Fig. B.42 Comparison of predictions made using Magnusson and Thelandersson's method (Type C) to data from Cardington test #5

made here. A comparison of burning rate predictions using Babrauskas' method to the CIB data for ventilation-controlled fires is presented in Fig. B.57.

The closed-form approximation was used to create predictions of compartment fire temperatures for the Cardington tests. In these tests, it was apparent that the fires

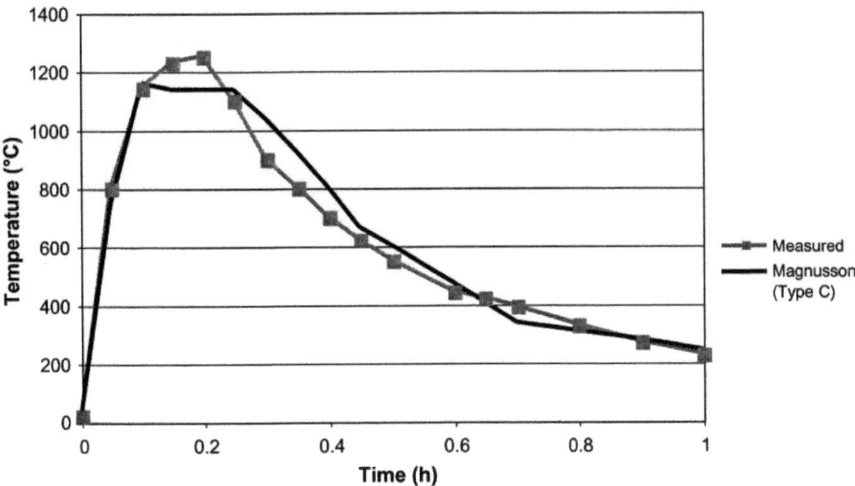

Fig. **B.43** Comparison of predictions made using Magnusson and Thelandersson's method (Type C) to data from Cardington test #7

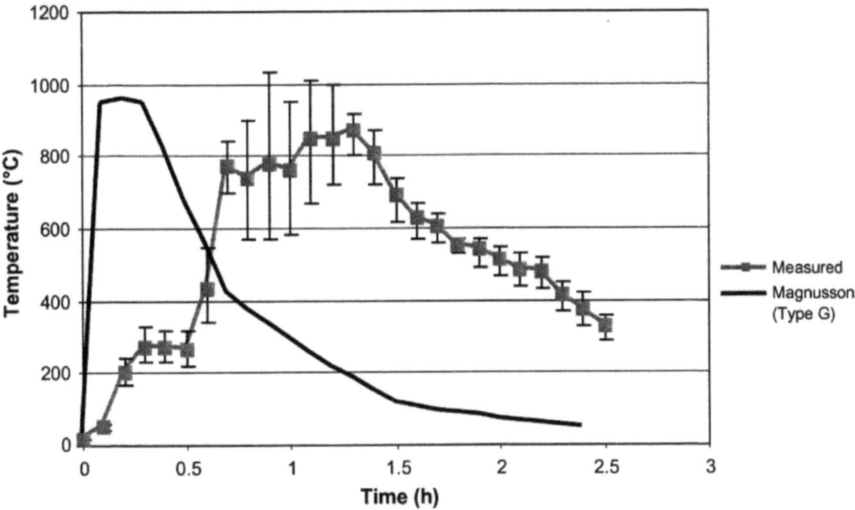

Fig. **B.44** Comparison of predictions made using Magnusson and Thelandersson's method (Type C) to data from Cardington test #8

were ventilation-controlled from the observed burning behavior. While Babrauskas' method is capable of predicting the burning rate and compartment fire temperatures during the growth and decay stages of a fire, these stages were neglected. The burning rate was calculated as [45]:

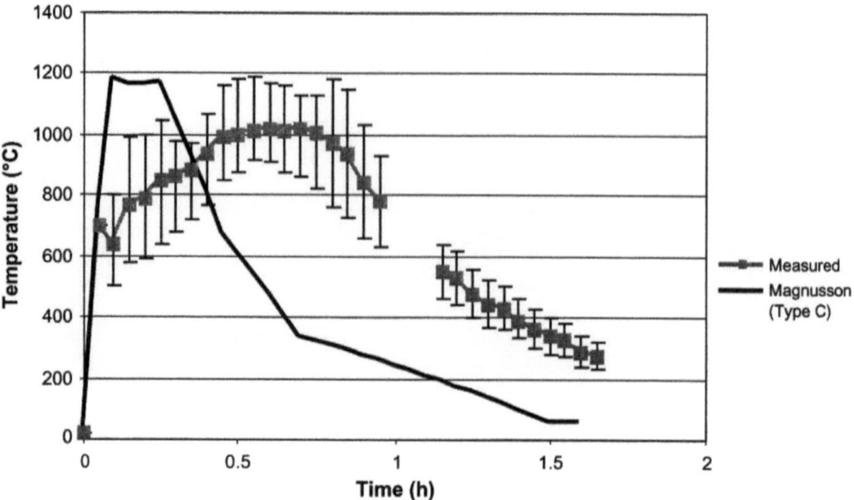

Fig. B.45 Comparison of predictions made using Magnusson and Thelandersson's method (Type C) to data from Cardington test #9

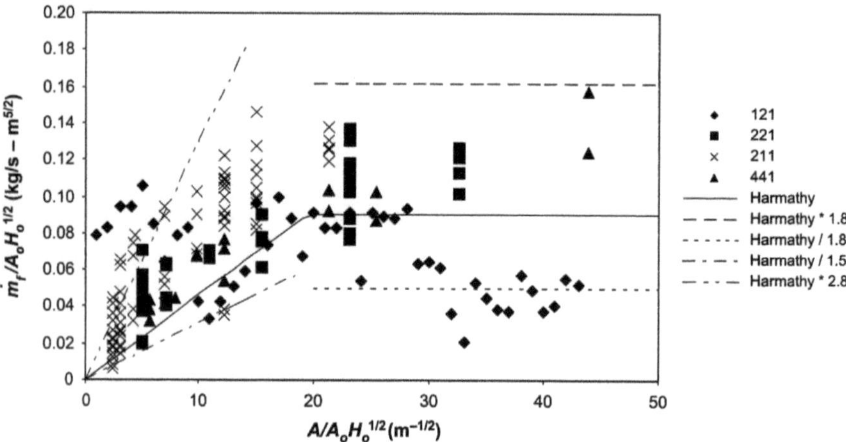

Fig. B.46 Comparison of CIB burning rate data to predictions made using Harmathy's method

$$\dot{m}_f = 0.12 A_o \sqrt{H_o}$$

Once the fuel was depleted, the fire was considered to cease, and the temperature was assumed to immediately return to ambient. Thus, the only time-dependent variable remaining was θ_3, which very quickly equaled one. Therefore, compartment fire temperatures were modeled as a square wave.

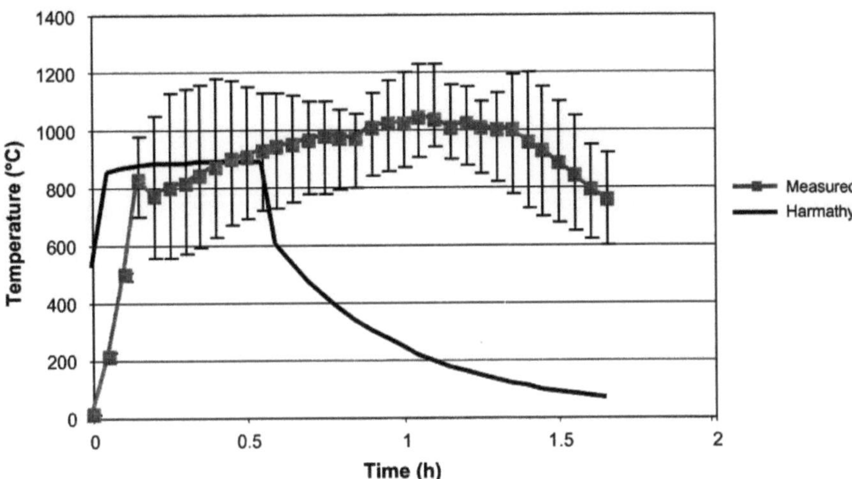

Fig. B.47 Comparison of predictions made using Harmathy's method to data from Cardington test #1

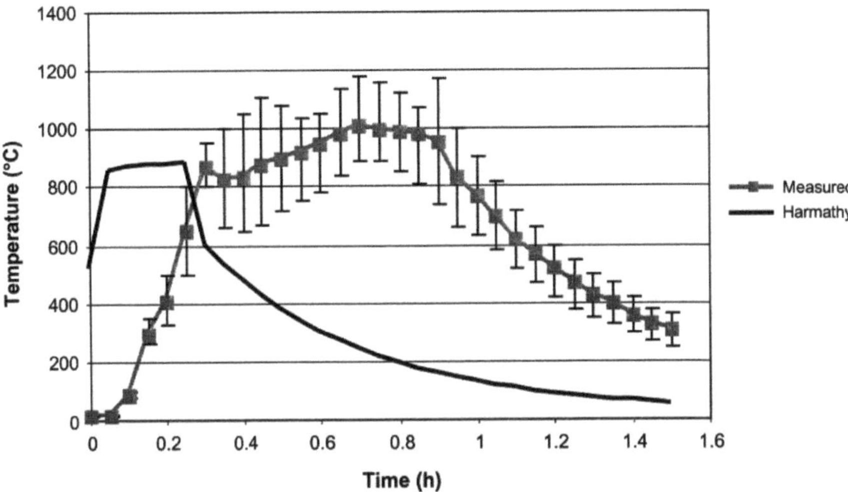

Fig. B.48 Comparison of predictions made using Harmathy's method to data from Cardington test #2

The value of s was calculated as 6.0, based on the chemical formula for typical wood provided by Harmathy [39] of $CH_{1.455}O_{0.645} \cdot 0.233H_2O$.

Calculations of the wall area did not include either the area of the floor or the area of the ventilation opening. The lining properties used were those of the ceramic fiber lining. For the calculation of θ_5, a value of 0.9 was used for b_p. The burning duration was calculated by dividing the mass of unburned fuel by the burning rate.

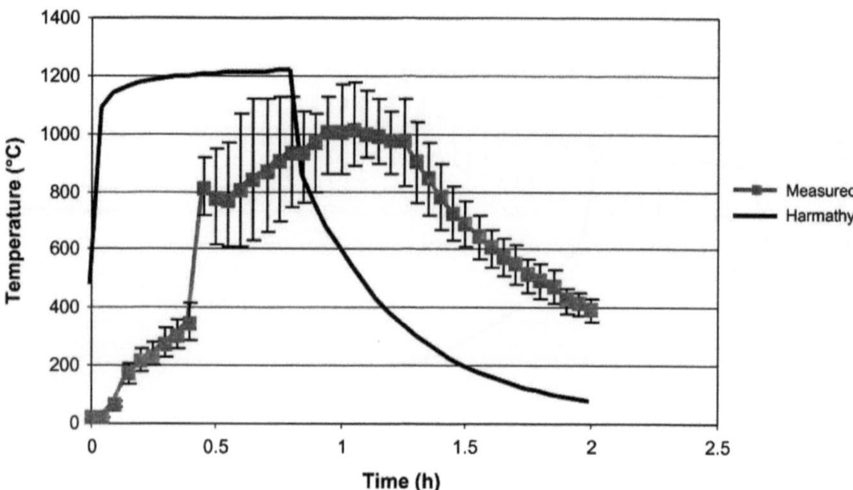

Fig. B.49 Comparison of predictions made using Harmathy's method to data from Cardington test #3

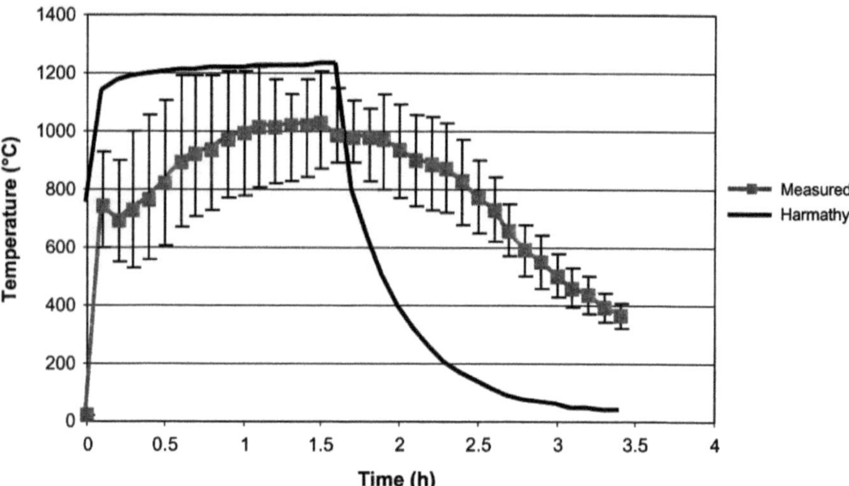

Fig. B.50 Comparison of predictions made using Harmathy's method to data from Cardington test #4

Comparisons of predictions using Babrauskas' method to the Cardington data are presented in Figs. B.58, B.59, B.60, B.61, B.62, B.63, B.64, B.65 and B.66.

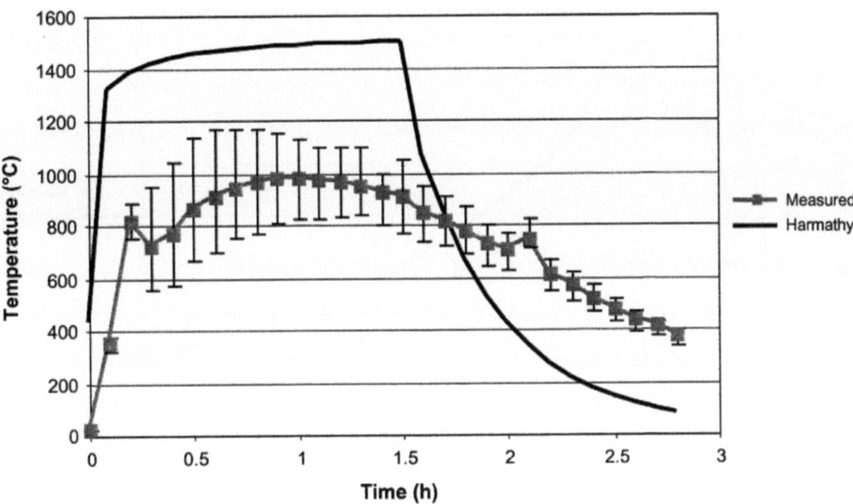

Fig. B.51 Comparison of predictions made using Harmathy's method to data from Cardington test #5

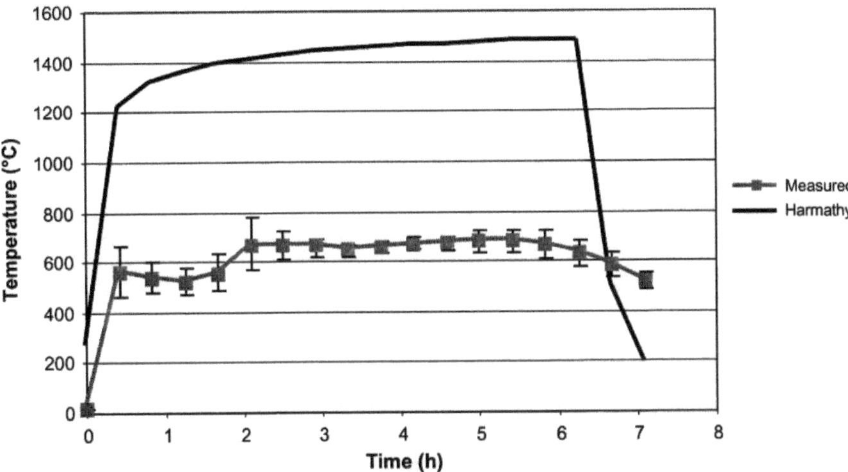

Fig. B.52 Comparison of predictions made using Harmathy's method to data from Cardington test #6

Ma and Mäkeläinen

Ma and Mäkeläinen define the critical value of $\frac{A}{A_o\sqrt{H_o}}$ that separates the fuel-controlled and ventilation-controlled regimes as

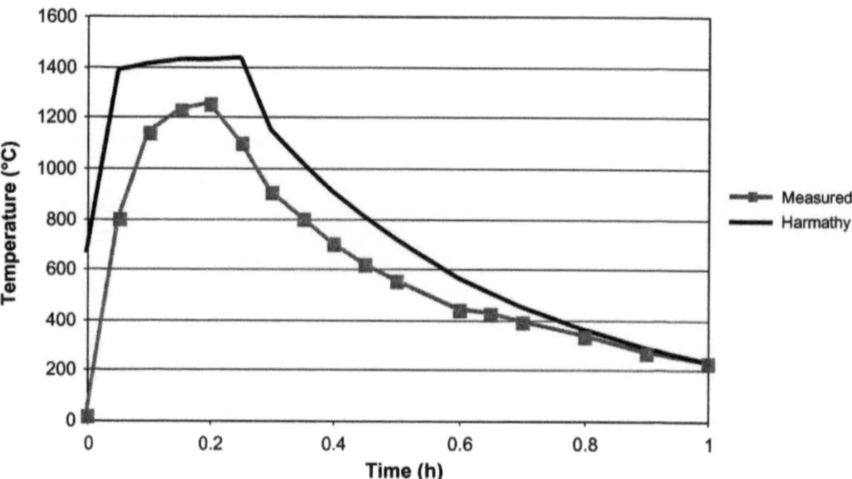

Fig. B.53 Comparison of predictions made using Harmathy's method to data from Cardington test #7

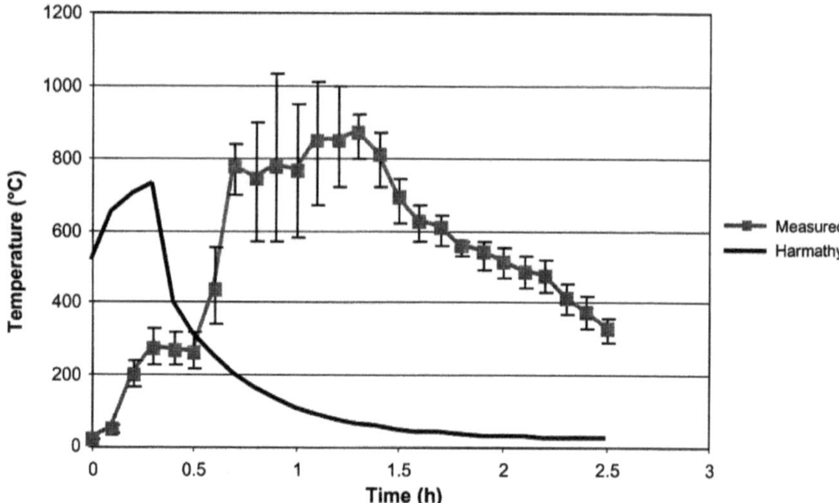

Fig. B.54 Comparison of predictions made using Harmathy's method to data from Cardington test #8

$$\frac{A_o\sqrt{H_o}}{A} = 0.07\,\frac{A_{\text{floor}}}{A}\,m_f''\,\frac{A_f}{m_f}$$

In the CIB tests, the ratio A_{floor}/A ranged from 0.18 to 0.25. Ma and Mäkeläinen noted that A_f/m_f typically ranges from 0.1 to 0.4 m^2/kg and that in a series of

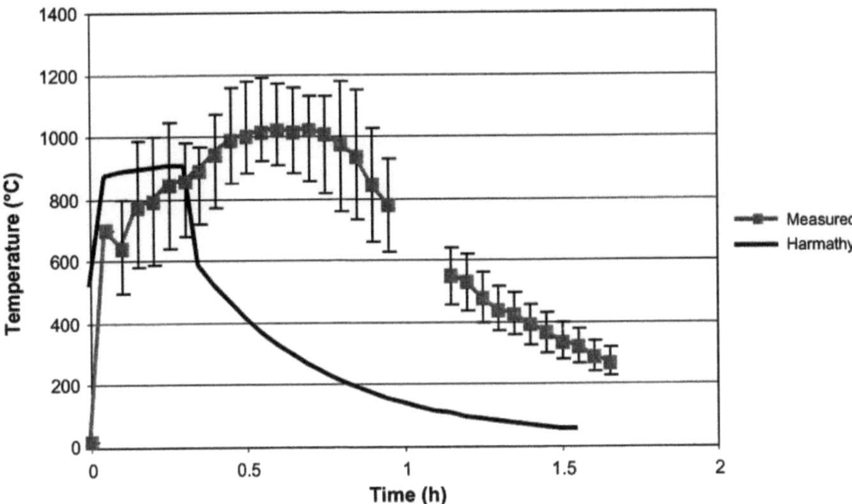

Fig. B.55 Comparison of predictions made using Harmathy's method to data from Cardington test #9

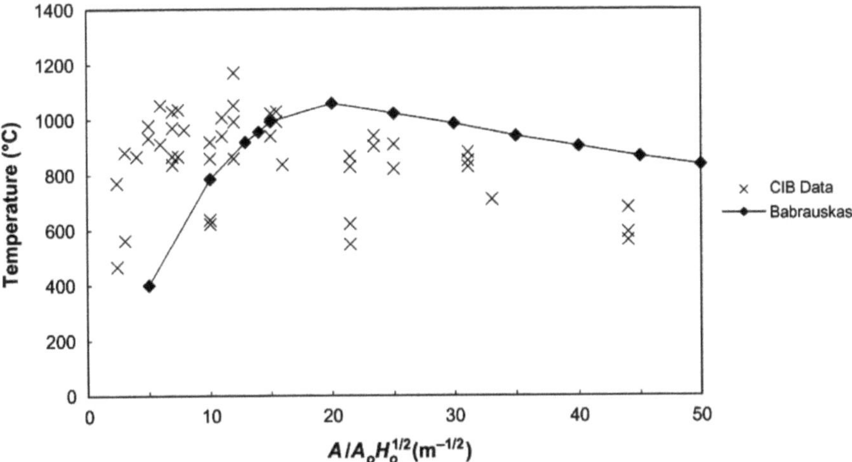

Fig. B.56 Comparison of CIB temperature data to predictions made using Babrauskas' method

Japanese tests, $A_f/m_f = 0.131$ m^2/kg. Substituting $A_{\text{floor}}/A = 0.2$, $A_f/m_f - 0.131$ m^2/kg, and $m''_f = 40$ kg/m^2, the critical value of $\frac{A_o\sqrt{H_o}}{A}$ that separates the fuel-controlled and ventilation-controlled regimes would be $\frac{A_o}{A_o\sqrt{H_o}} = 13.68$.

Ma and Mäkeläinen estimate the maximum temperature that would be achieved for ventilation-controlled fires would be:

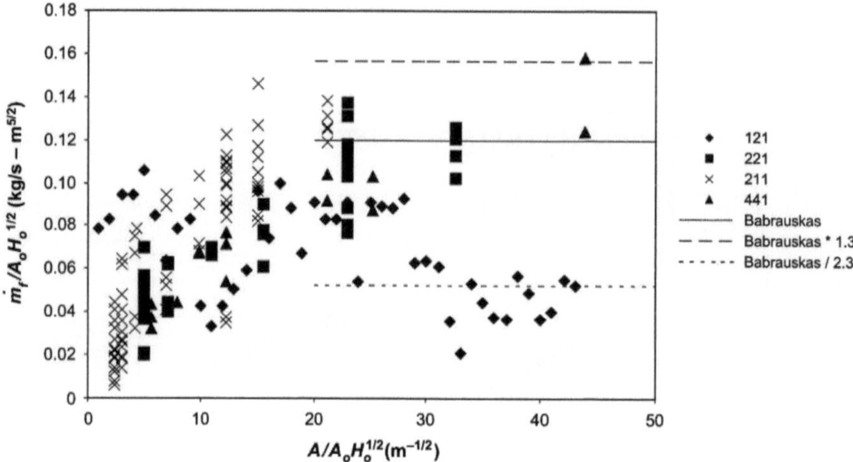

Fig. B.57 Comparison of CIB burning rate data to predictions made using Babrauskas' Method

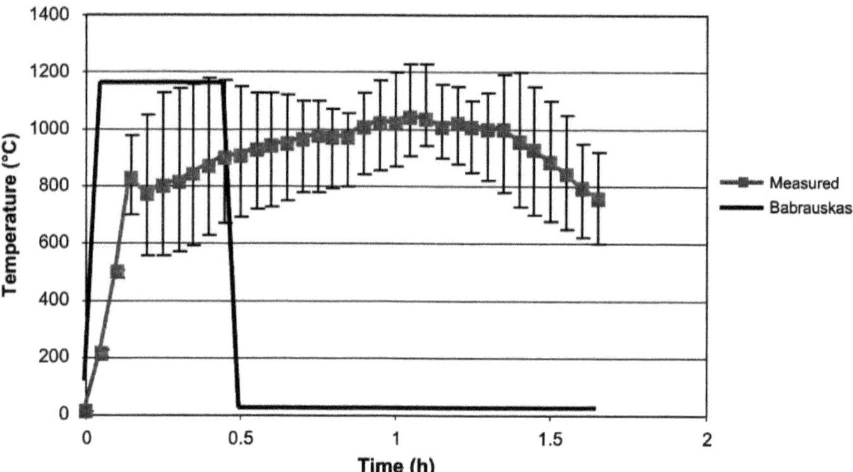

Fig. B.58 Comparison of predictions made using Babrauskas' method to data from Cardington test #1

$$T_{gm} = 1240 - 11\frac{A}{A_o\sqrt{H_o}} \;\; (^{\circ}\text{C})$$

For fuel-controlled fires, Ma and Mäkeläinen state that the maximum temperature would be $T_{gm} = T_{gmcr}\sqrt{\frac{A}{\eta_{cr}A_o\sqrt{H_o}}}$, where η_{cr} is the value of $\frac{A_o}{A_o\sqrt{H_o}}$ that differentiates between fuel- and ventilation-controlled burning (for the CIB data, η_{cr} was calculated as 13.68 m$^{-1/2}$), and T_{gmcr} is the value of T_{gm} for $\eta = \eta_{cr}$. It should be noted that

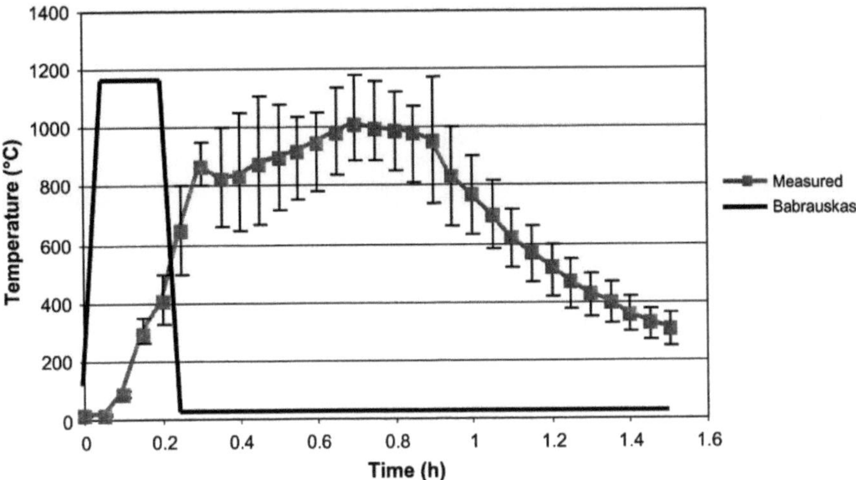

Fig. B.59 Comparison of predictions made using Babrauskas' method to data from Cardington test #2

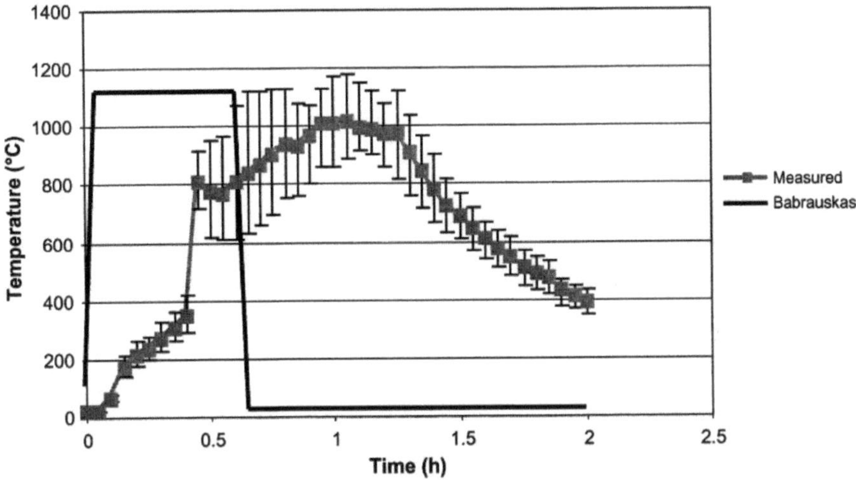

Fig. B.60 Comparison of predictions made using Babrauskas' method to data from Cardington test #3

the above temperature correlations provide an estimation of the maximum temperature that would be attained during a fire; for the majority of the fire duration, the temperature would be lower, and, hence, die average temperature during the fire would be lower. Figure B.67 provides a comparison of predicted maximum temperatures with the CIB data.

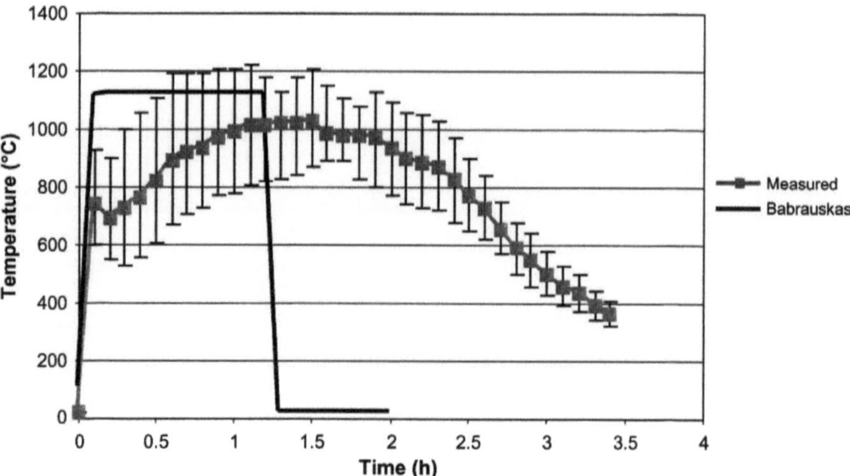

Fig. B.61 Comparison of predictions made using Babrauskas' method to data from Cardington test #4

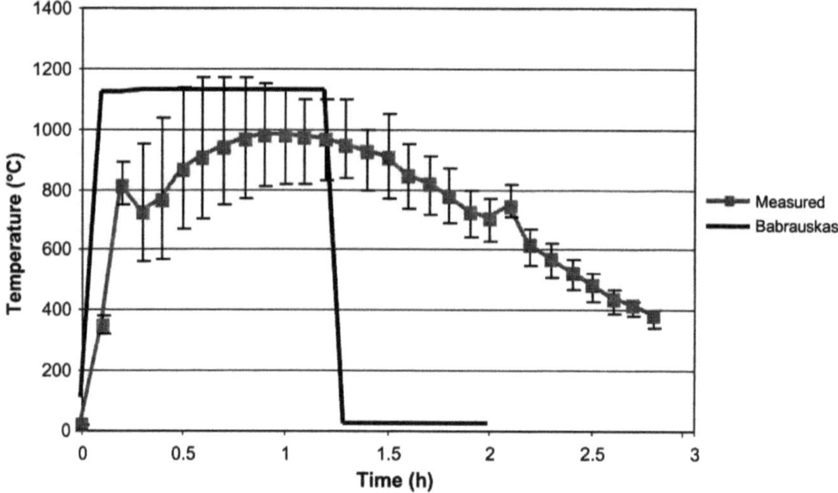

Fig. B.62 Comparison of predictions made using Babrauskas' method to data from Cardington test #5

Ma and Mäkeläinen use Harmathy's correlation to predict the burning rate for fuel-controlled burning and Law's correlation to predict the burning rate for ventilation-controlled burning. See the discussion of those methods for an evaluation of their burning rate predictions.

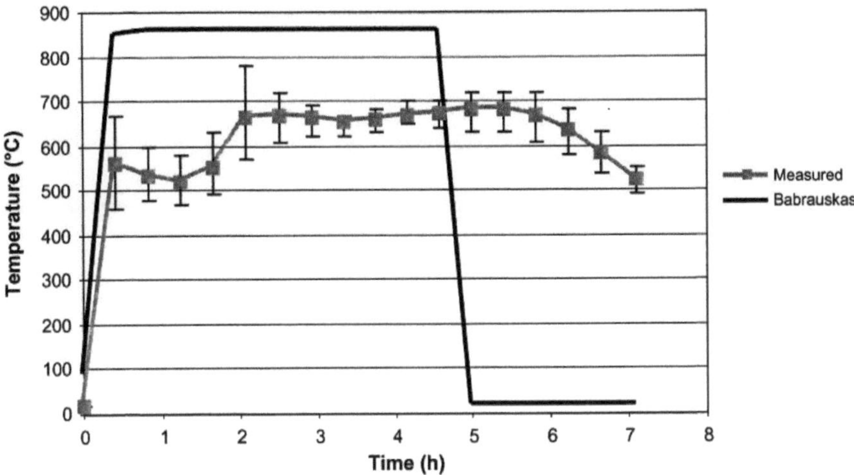

Fig. B.63 Comparison of predictions made using Babrauskas' method to data from Cardington test #6

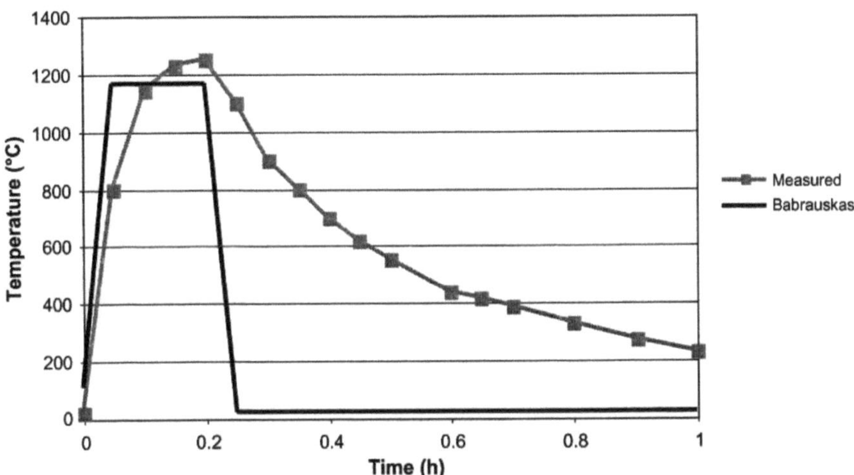

Fig. B.64 Comparison of predictions made using Babrauskas' method to data from Cardington test #7

Comparisons of predictions to the Cardington data are presented in Figs. B.68, B.69, B.70, B.71, B.72, B.73, B.74 and B.75. For test #6, Ma and Mäkeläinen's method predicted temperatures below ambient.

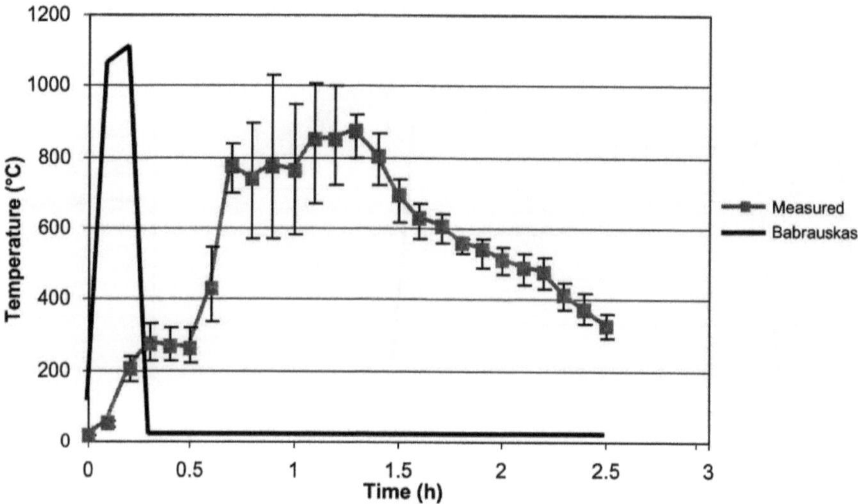

Fig. B.65 Comparison of predictions made using Babrauskas' method to data from Cardington test #8

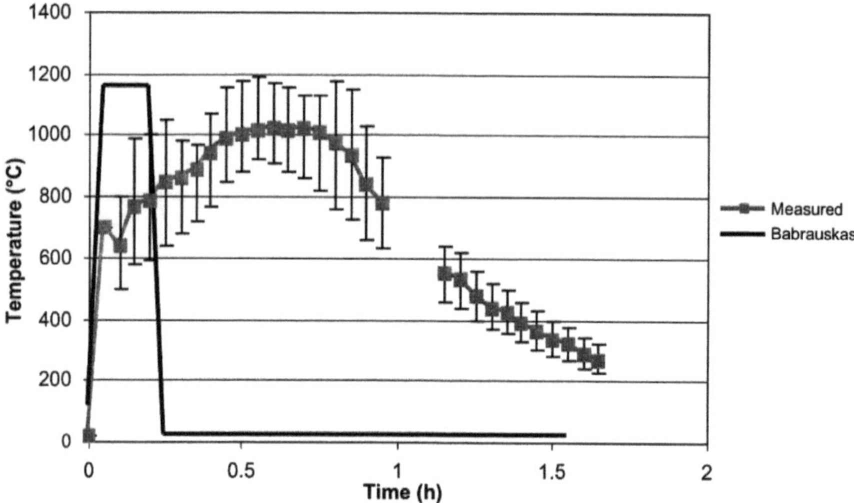

Fig. B.66 Comparison of predictions made using Babrauskas' method to data from Cardington test #9

CIB

The temperature data from the Cardington tests was compared to the temperature data from the CIB tests by averaging the temperatures measured at different horizontal locations in the Cardington tests. These average temperatures were averaged

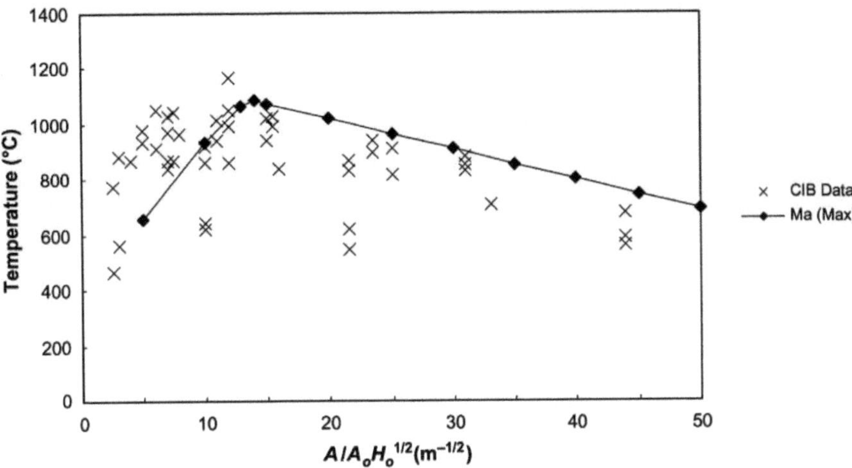

Fig. B.67 Comparison of CIB burning rate data to predictions made using Ma and Mäkeläinen's method

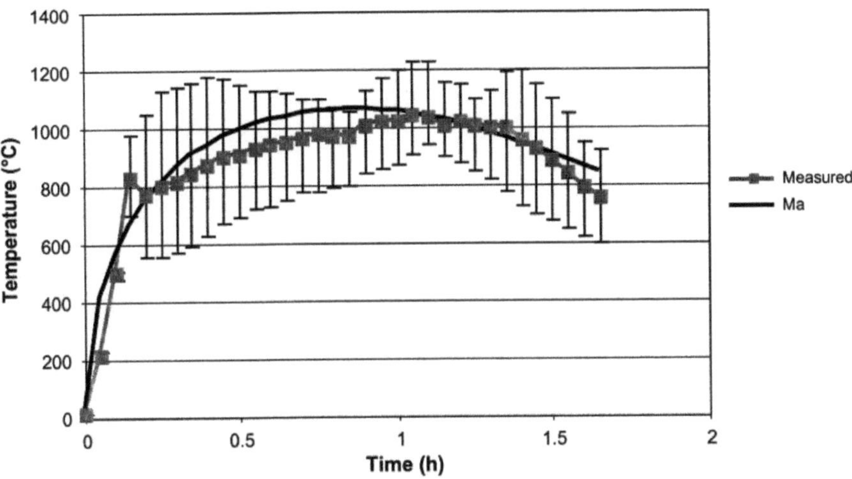

Fig. B.68 Comparison of predictions made using Ma and Mäkeläinen's method to data from Cardington test #1

over the duration of maximum burning and plotted along with the CIB data. Error bars on the Cardington data are included to show the range of temperatures measured during the period of maximum burning. The results are shown in Fig. B.76, with the abscissa plotted in a logarithmic scale.

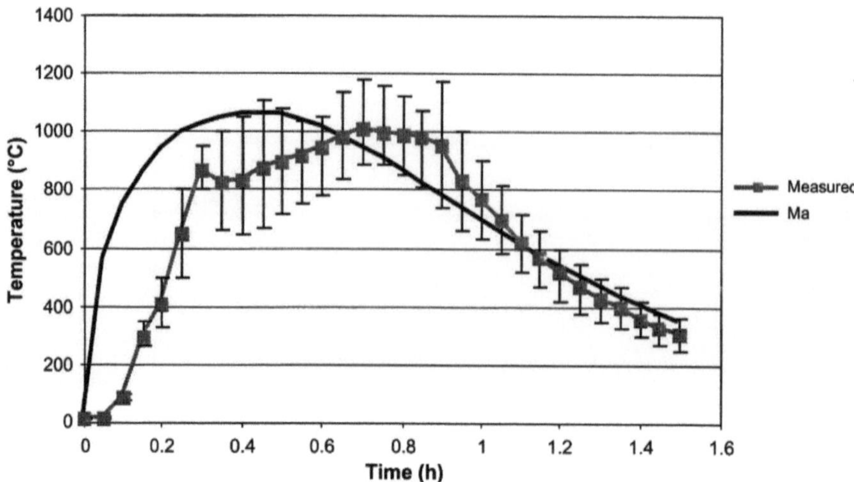

Fig. B.69 Comparison of predictions made using Ma and Mäkeläinen's method to data from Cardington test #2

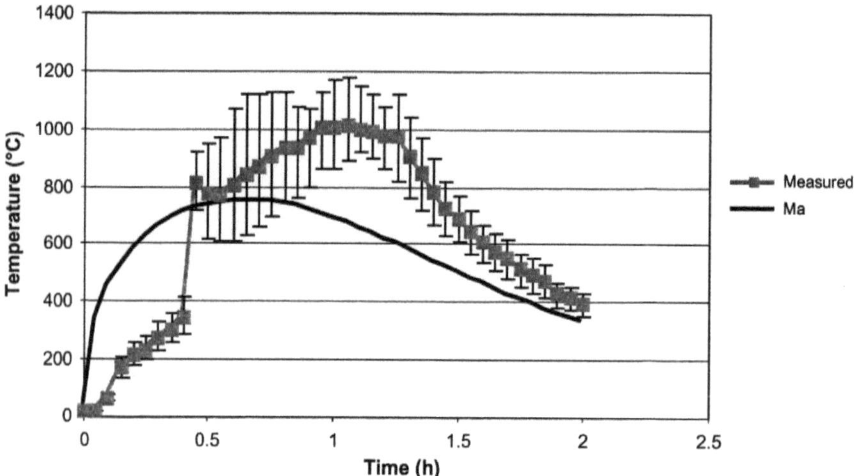

Fig. B.70 Comparison of predictions made using Ma and Mäkeläinen's method to data from Cardington test #3

Predictions using the CIB method are compared to data from the Cardington tests in Figs. B.77, B.78, B.79, B.80, B.81, B.82 and B.83. The compartment temperature and burning duration were predicted using the graphs presented earlier in this guide for cribs with 20 mm thick wood sticks spaced 20 mm apart. No decay rate was imposed, and for times greater than the duration, the compartment temperature was assumed to be ambient.

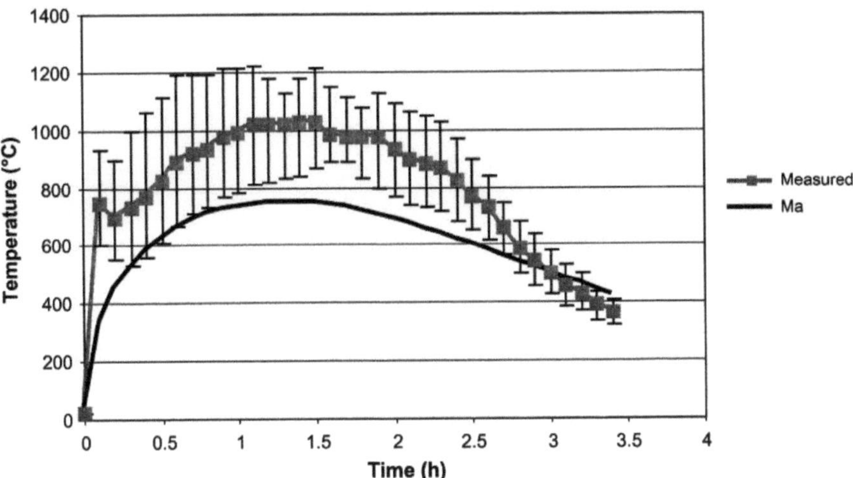

Fig. B.71 Comparison of predictions made using Ma and Mäkeläinen's method to data from Cardington test #4

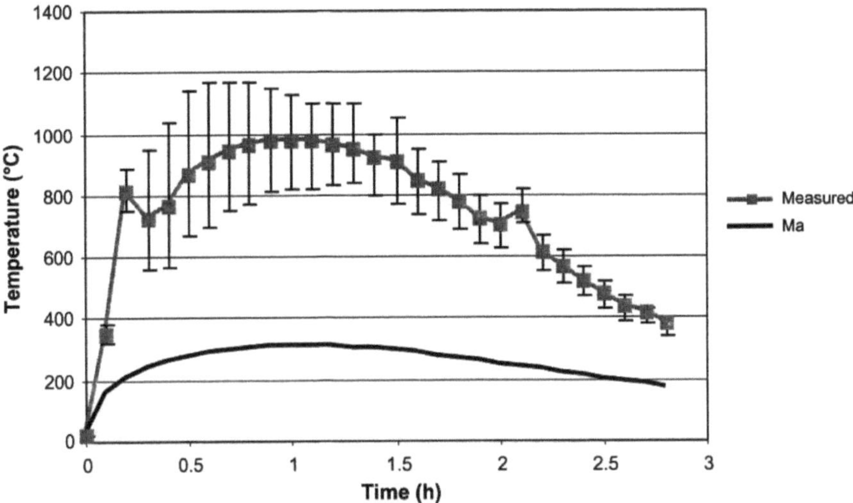

Fig. B.72 Comparison of predictions made using Ma and Mäkeläinen's method to data from Cardington test #5

Law

Figure B.84 shows predictions of maximum temperature using Law's method compared to the CIB data. Law's method includes a means of reducing the predicted temperature based on the fuel loading. However, for the range of conditions in the tests from which the CIB data were collected, utilizing this factor would result in

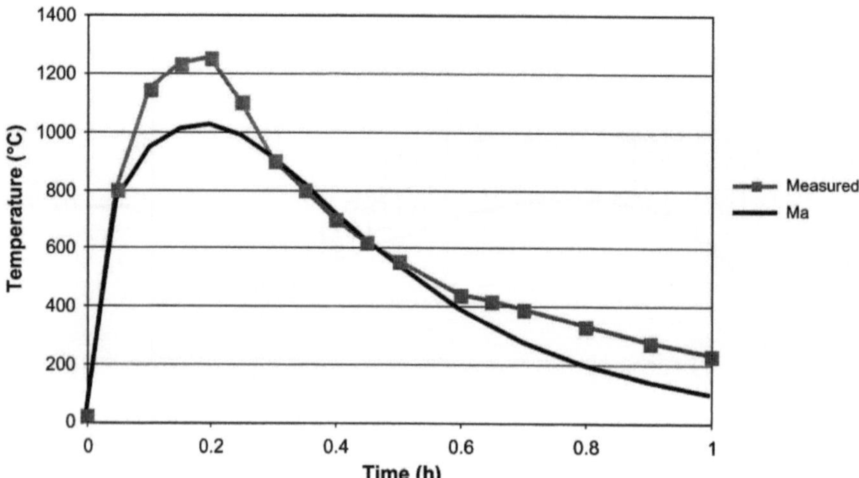

Fig. B.73 Comparison of predictions made using Ma and Mäkeläinen's method to data from Cardington test #7

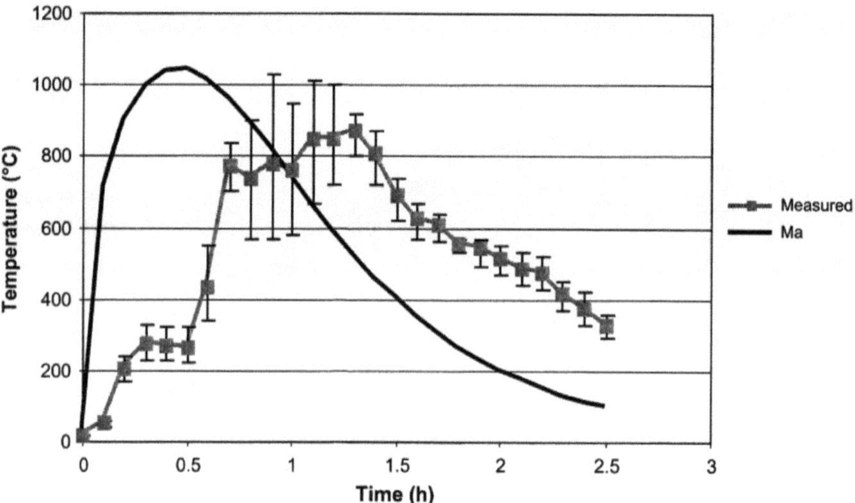

Fig. B.74 Comparison of predictions made using Ma and Mäkeläinen's method to data from Cardington test #8

unrealistically low temperatures for some combinations of scale, opening factor, and ventilation area. Therefore, this method of reducing the temperature was not utilized.

Figure B.85 shows a comparison of burning rate predictions made using Law's method to the CIB data. Note that, because Law's method considers the effect of compartment depth and width, the CIB burning rate data that was normalized by $\sqrt{D/W}$ was utilized.

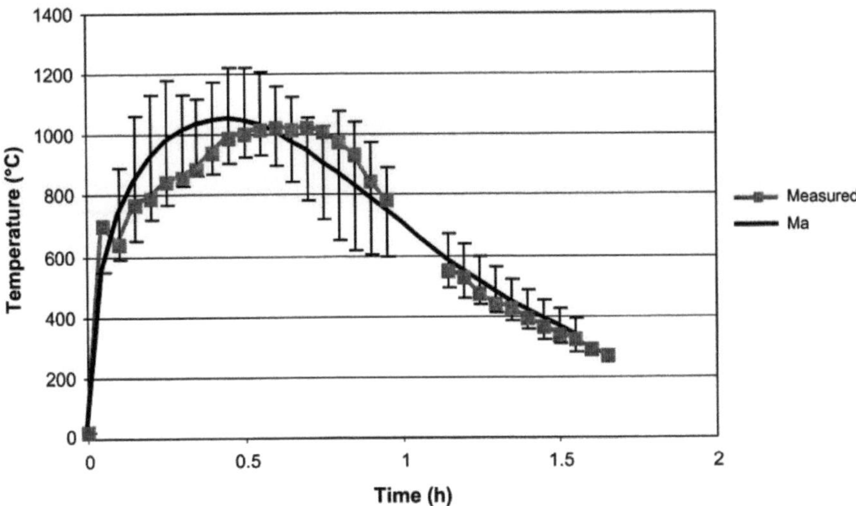

Fig. B.75 Comparison of predictions made using Ma and Mäkeläinen's method to data from Cardington test #9

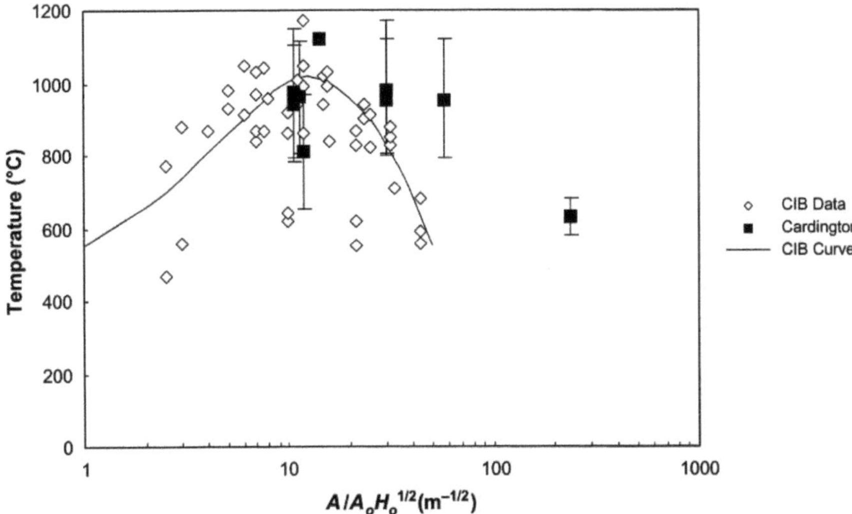

Fig. B.76 Comparison of Cardington and CIB temperature data

Comparisons of predictions made using Law's method to the Cardington data are shown in Figs. B.86, B.87, B.88, B.89, B.90, B.91, B.92, B.93 and B.94. For times less than the calculated burning duration, the temperature was calculated using Law's adjustment for fuel load. No decay rate was imposed, and for times greater than the duration, the compartment temperature was assumed to be ambient.

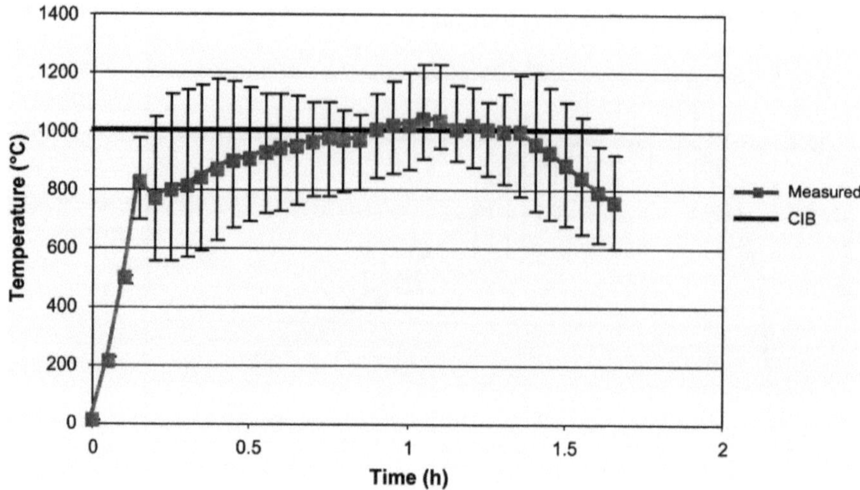

Fig. B.77 Comparison of predictions made using the CIB data to Cardington test #1

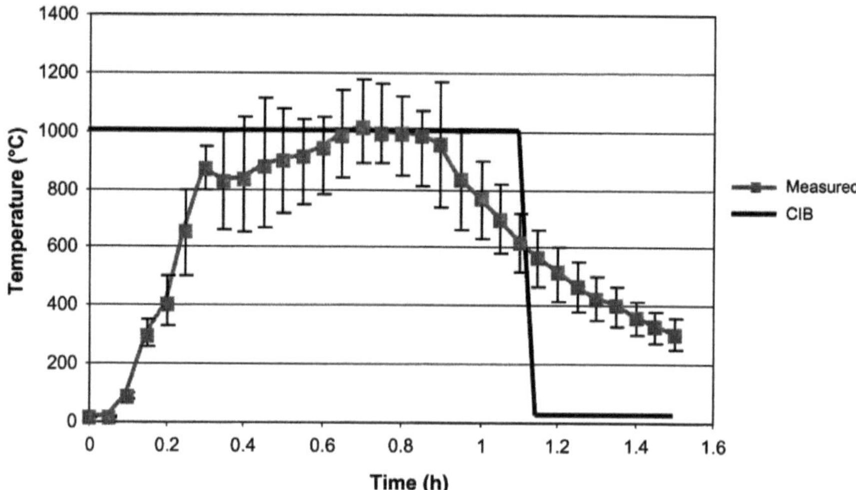

Fig. B.78 Comparison of predictions made using the CIB data to Cardington test #2

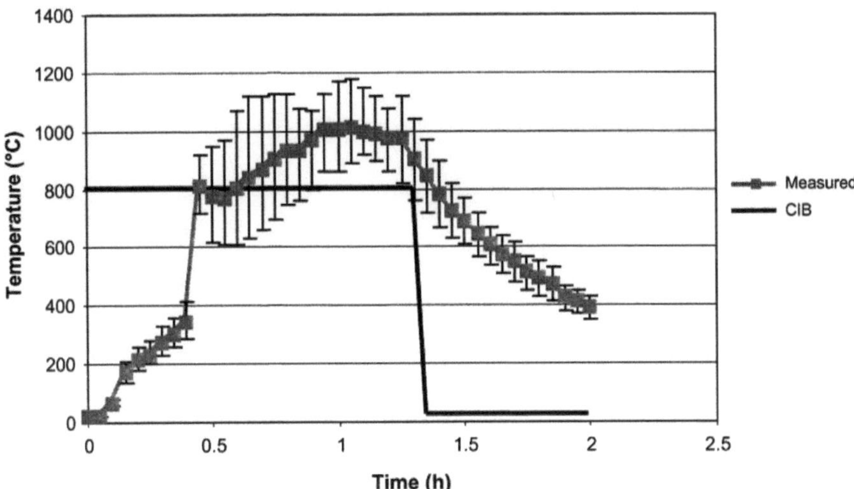

Fig. B.79 Comparison of predictions made using the CIB data to Cardington test #3

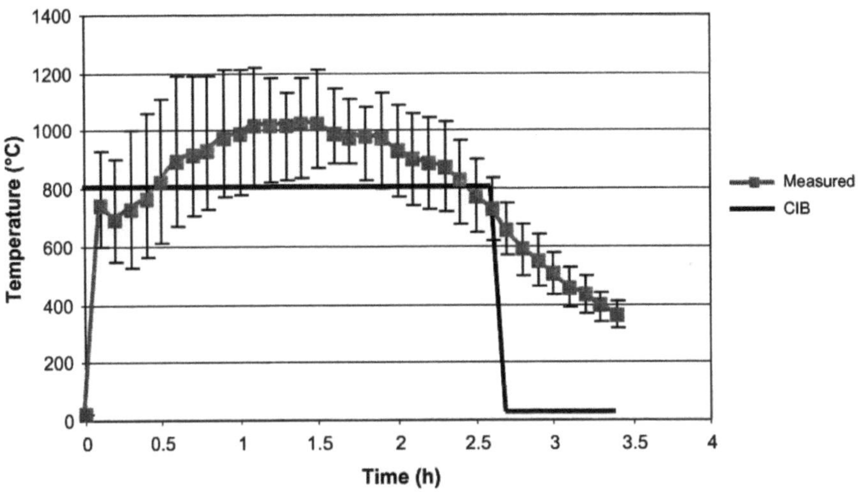

Fig. B.80 Comparison of predictions made using the CIB data to Cardington test #4

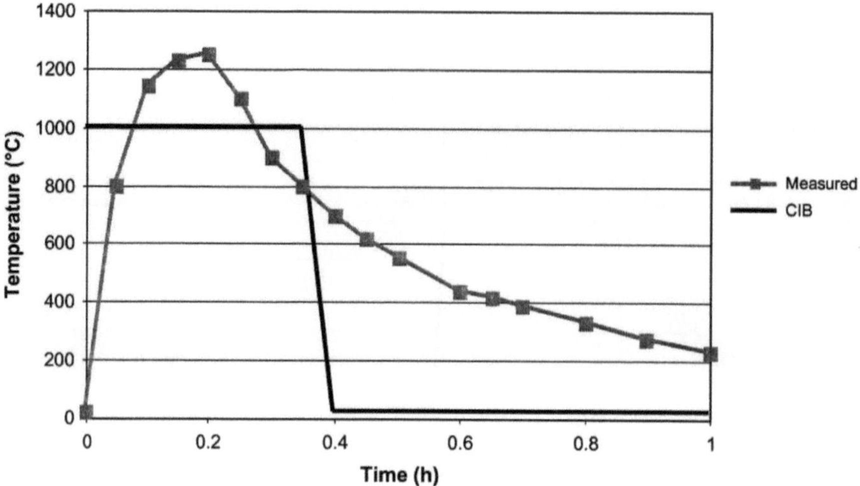

Fig. B.81 Comparison of predictions made using the CIB data to Cardington test #7

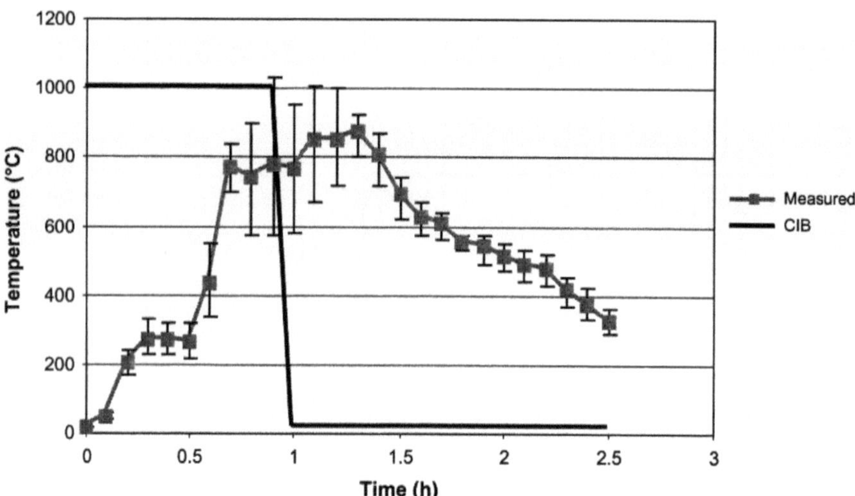

Fig. B.82 Comparison of predictions made using the CIB data to Cardington test #8

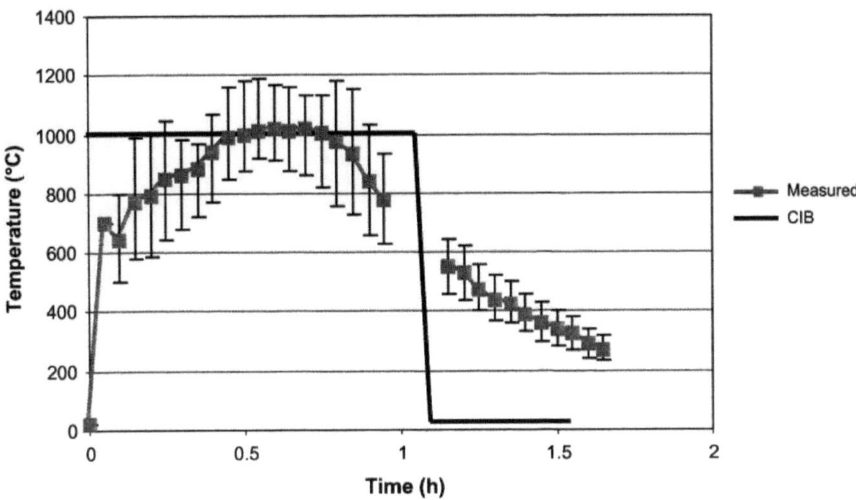

Fig. B.83 Comparison of predictions made using the CIB data to Cardington test #9

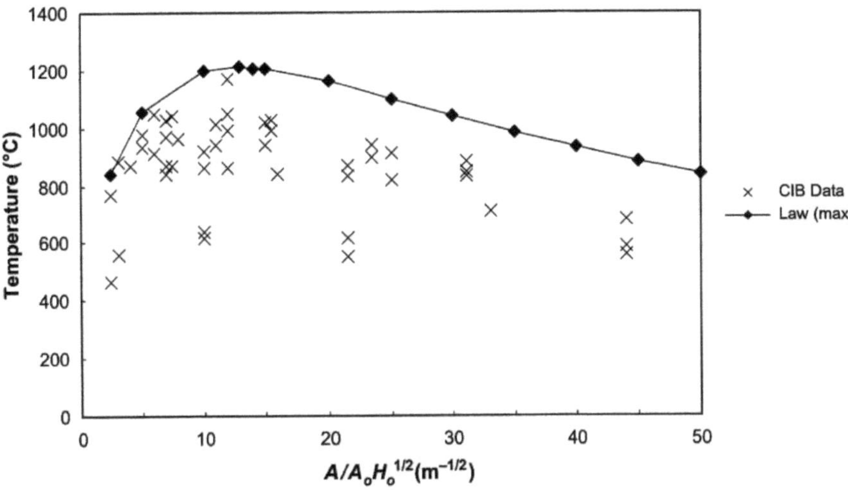

Fig. B.84 Comparison of CIB temperature data to predictions made using Law's method

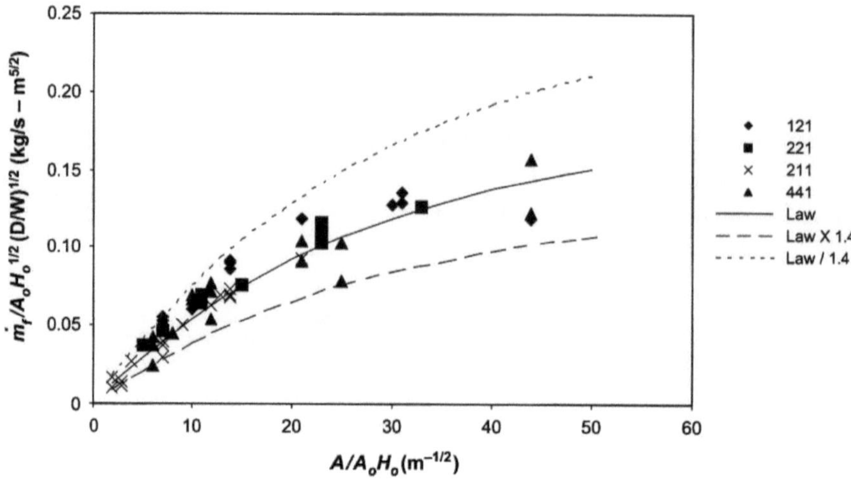

Fig. B.85 Comparison of CIB burning rate data to predictions made using Law's method

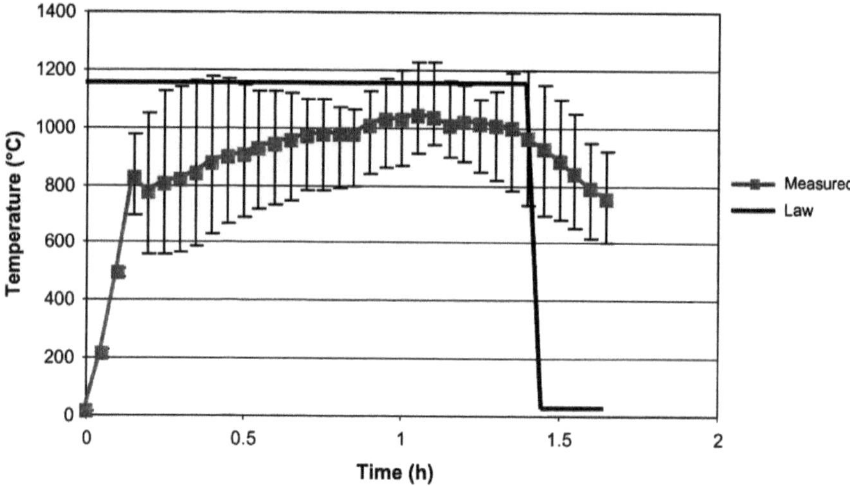

Fig. B.86 Comparison of predictions made using Law's method to data from Cardington test #1

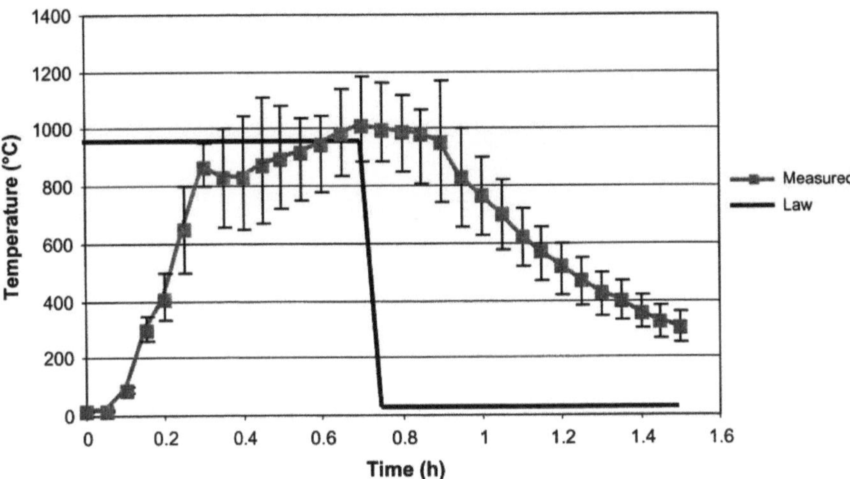

Fig. B.87 Comparison of predictions made using Law's method to data from Cardington test #2

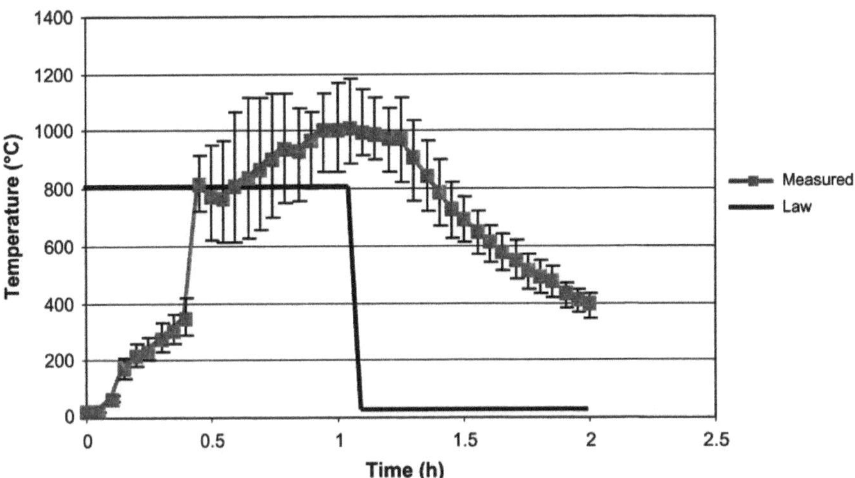

Fig. B.88 Comparison of predictions made using Law's method to data from Cardington test #3

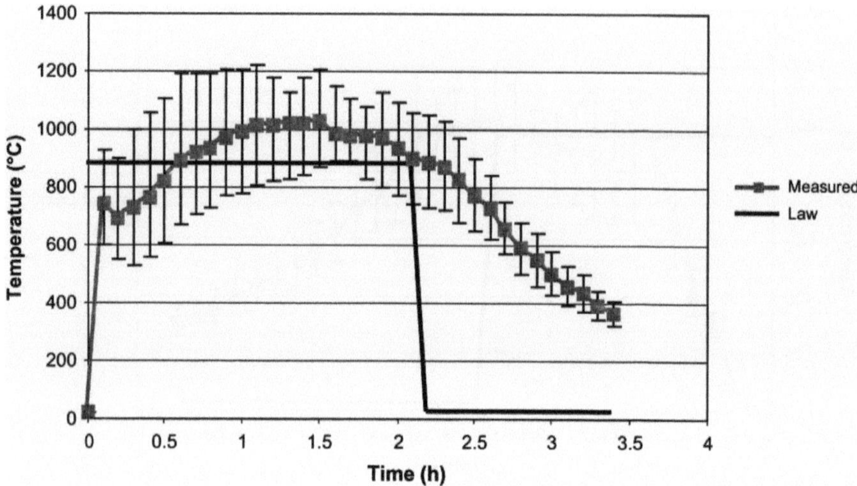

Fig. B.89 Comparison of predictions made using Law's method to data from Cardington test #4

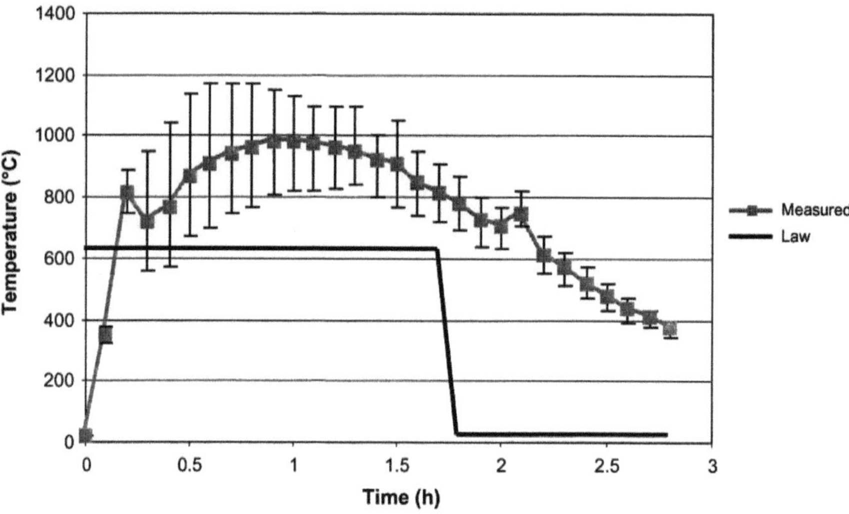

Fig. B.90 Comparison of predictions made using Law's method to data from Cardington test #5

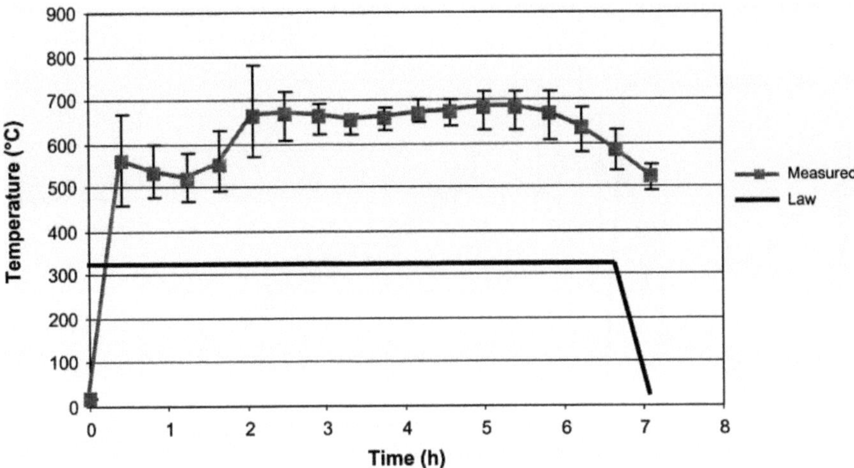

Fig. B.91 Comparison of predictions made using Law's method to data from Cardington test #6

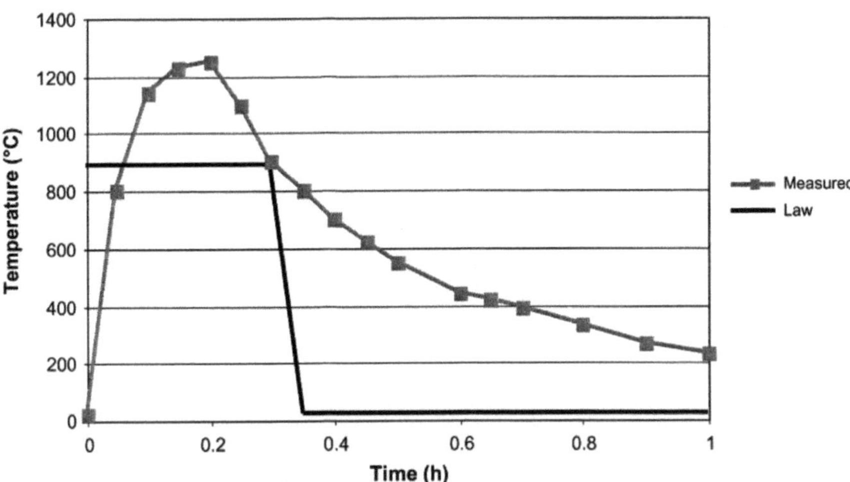

Fig. B.92 Comparison of predictions made using Law's method to data from Cardington test #7

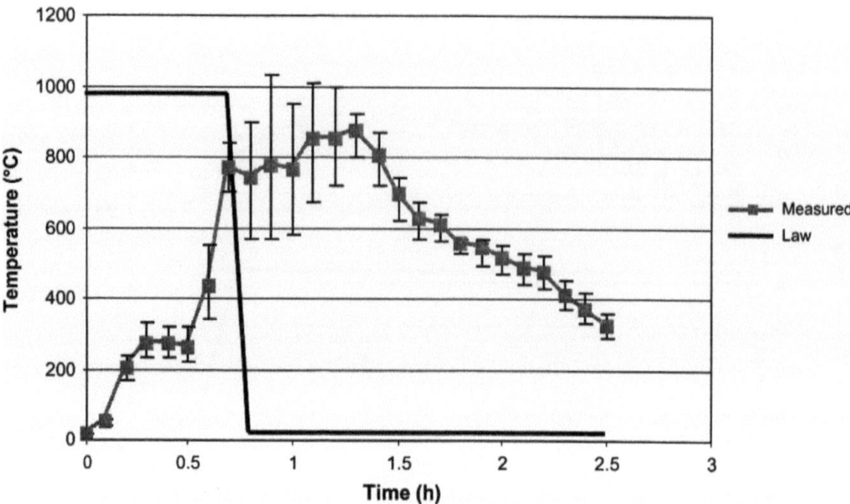

Fig. B.93 Comparison of predictions made using Law's method to data from Cardington test #8

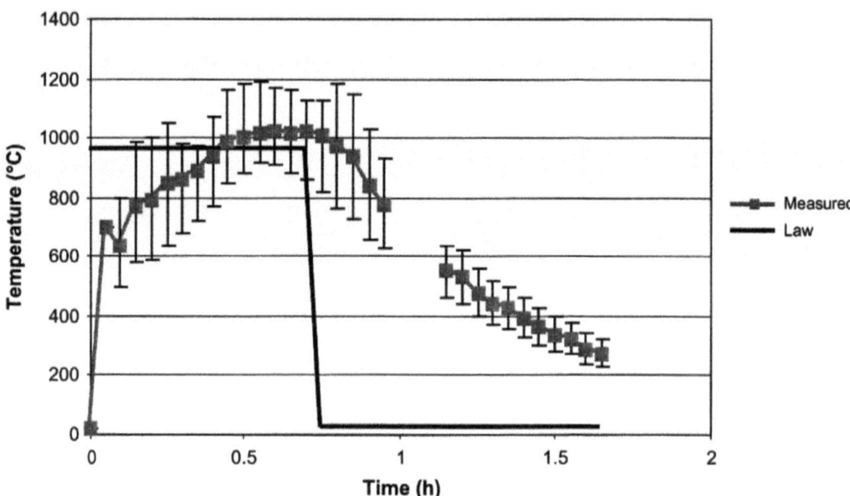

Fig. B.94 Comparison of predictions made using Law's method to data from Cardington test #9

Appendix C: Time-Equivalent Methods

As stated in ASTM E119 [105], standard furnace tests such as ASTME119; BS 476, Part 20 [106]; and ISO 834 [107] provide a relative measure of the fire test response of comparable assemblies under standardized fire exposure conditions. The exposure is not representative of all fire conditions because conditions vary with changes in the amount, nature, and distribution of fire loading, ventilation, compartment size and configuration, and thermal characteristics of the compartment. Real fires can be more or less severe in terms of duration, rate of heating, and peak temperature than the standard temperature-time relationship in a furnace test.

Real fires are a function of fuel load, compartment dimensions, thermal properties of the compartment boundaries, and the quantity of unprotected openings that allow ventilation in a post-flashover fire.

Also of importance is that the standard furnace test does not assess real structural response in fire conditions because single elements of structure are tested in the furnace even though they form component parts of complex three-dimensional flames in real buildings.

Various methods exist for designers to derive more realistic temperature–time relationships for compartments. For example, as a result of concerns with the standard furnace test temperature–time relationship, work was carried out by Ingberg, Law, and Pettersson, among others, to determine what is known as an equivalent fire resistance. For these methods, the heating effect in a compartment is based on real compartment fire behavior and therefore takes into account fuel load density, ventilation openings, compartment dimensions, and enclosure thermal properties. This allows some improvement in the grading method based on the standard furnace test that is currently assumed in building codes worldwide.

This section describes various calculation procedures for these time-equivalent methods. Limitations and assumptions for each method are described. Pettersson's method is put forward as the preferred time-equivalent method, and its range of use is outlined.

Real Structural Response

It is important to note that time-equivalent methods do not assess local or global structural response. They relate only to heating effects and their relationship to the standard furnace test.

The t-equivalent methods do not address transient temperature gradients or associated load-bearing capacities. The ratings derived do not relate to actual frame performance in fire. These methods are simply refined versions of the performance of a single element in fire, but only relative to the standard furnace test. They normally assume insulated structures only (protected steel or reinforced concrete). Pettersson's work, however, does address uninsulated steel also. In the work carried out for the Natural Fire Safety Concept [108], good correlations were achieved when the t-equivalent results were compared to real fire test data for insulated steel structures. The results for uninsulated steel structures gave a very poor correlation, as would be expected. Bare steel tends to follow the furnace test curve, so the use of bare steel elements, in terms of standard fire resistance, would not be expected beyond 20–30 min depending on section size.

A time-equivalent calculation does not apply if the pre-flashover calculations show that the flashover will not occur, i.e., the calculation is no longer relevant if the flashover has not occurred. Then, local heating effects are relevant, not temperatures in a uniformly heated compartment, as is assumed in time-equivalent analysis methods.

Time-equivalent methods are empirical formulae developed by regression analysis using a selected number of tests or calculations. Therefore, they have been developed for a certain range of structural steel sizes and thicknesses of insulation, and so may not be appropriate outside this range.

They are used for other materials, but beyond protected steel and reinforced concrete, very little is known of the accuracy of applying this method to other materials.

Note that all t-equivalent methods described here involve combustible solids only.

Discussion of Methods

Time-equivalent methods can be described as methods that define the thermal exposure of a particular compartment fire in terms of the duration of the equivalent standard fire.

Equivalence of thermal exposure has been defined in two ways:

1. Equal areas under the temperature–tune curves
2. Equal temperatures at the critical part of a structural element

The two methods give similar results where the element selected has a fire resistance of the order of half an hour or more.

Fire Load Concept

By 1918, there was a concern in the fire protection and code enforcement communities that there was no accepted method for establishing appropriate levels of fire endurance for buildings of different sizes and occupancies [109]. The original work had been based on "fireproof' large commercial buildings. It was recognized that these differed significantly from residential fires, but it was not understood how their severity related to the conditions in the now-formulated standard fire resistance test.

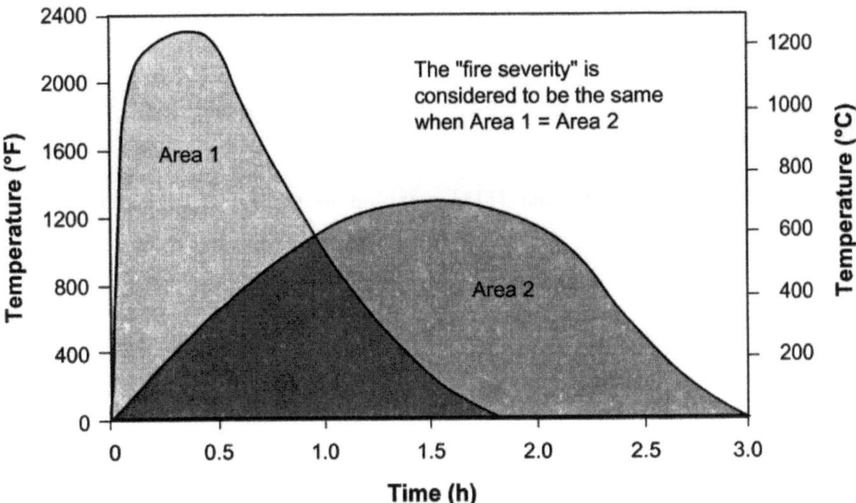

Fig. C.1 Fire severity concept [109]

To develop a solution to this problem, in 1922, the National Bureau of Standards investigated the nature of building fires under the direction of Simon Ingberg [110]. The main aim was to determine the title intensity and duration of uncontrolled fires in particular occupancies resulting from different levels of fuel load. Ingberg also investigated the validity of the standard temperature–time curve.

Ingberg investigated office- and record storage-type occupancies. The effects of the building size and fuel load, combustible and noncombustible flooring, plus wood and steel furniture were investigated.

As a result of these tests, Ingberg established a simple relationship between the average weight of combustible material within a room and the fire endurance necessary to withstand a complete burnout of the contents. This is known as the "fuel load concept." It assumes that the area under any temperature–time curve from ignition through decay provides a comparative measure of fire severity and that fire severity is a function of the fuel load only.

Ingberg compared the area under the temperature–time curves generated in the burnout tests to an equivalent area under the standard temperature-time curve. The areas below a threshold temperature of about 300 °C were not taken into account. The graph in Fig. C.1 shows the basis for Ingbergs [5] work.

Ingberg developed the following relationship for time-equivalence:

$$t_e = k_1 m''_f$$
(C.1)

where:

t_e = t-equivalent (min)
m''_f = Fuel load (wood) per unit floor area
k_1 = Unity when m''_f is in units of kg/m^2; k_1 = 5 when m''_f is in units of lb/ft^2

Ingberg's work became widely accepted as the general basis for establishing fire endurance requirements.

Kawagoe and Sekine

In 1963, Kawagoe and Sekine [111] went on to show the importance of the ventilation parameter:

$$\frac{A_o\sqrt{H_o}}{A}$$

where:

H_O = Window height (m)
A_O = Total area of openings (m^2)
A = Total area of inside surfaces including opening area (m^2)

Kawagoe and Sekine also developed a formula for fire duration and defined it as the period from the beginning of temperature rise until the time the temperature drops after most of the combustible material is burnt.

This time, τ is approximated as:

$$\tau = \frac{m_f''}{5.5}\frac{A_{\text{floor}}}{A_o\sqrt{H_o}} \ (\text{min}) \tag{C.2}$$

where:

τ = Time (min)
m_f'' = Fuel load (kg/m^2)
A_{floor} = Floor area (m^2)
H = Height of the window (m)
A_O = Area of the windows (m^2)

Law

Law developed a t-equivalent formula [112, 113] from the results of the CIB test program [114].

The maximum temperature that would be attained by a protected steel element in a real fire compartment was chosen as a basis for comparison with the heating effect in a standard fire.

For a temperature–time curve, the maximum temperature obtained by a protected steel element in a compartment fire is calculated as:

$$\frac{dT_s}{dt} = \frac{T - T_s}{RC} \tag{C.3}$$

where:

T_s = Steel temperature (K)
t = Time (s)
T = Fire temperature (K)
$R = \delta i \, / \, (k_i PH)$
δ_i = Thickness of insulating material (m)
k_i = Thermal conductivity of insulating material (kW/m-K)
P = Heated perimeter of steel member (m)
H = Height (or length) of steel member (m)
$C = AH\rho_s c_s$
A = Cross-sectional area of steel member (m^2)
ρ_s = Density of steel (kg/m^3)
c_s = Specific heat of steel (kJ/kg-K)

The temperature of the heated surface of the protective material is assumed to be the same as the fire temperature. The heat transfer through the steel section can then be calculated.

For a given temperature–time curve, the value RC was determined so that the maximum temperature of the protected member was 550 °C.

The time for the protected member to attain 550 °C when exposed to the standard temperature–time curve gives the value of t-equivalent.

The best correlation was obtained from the product (m_f/A_o) and a term taking into account A_o and the solid surface to which heat is lost:

$$t_e = k_3 m_f / (A_o [A - A_{\text{floor}} - A_o])^{1/2} \tag{C.4}$$

where:

A_{floor} = Floor area of the compartment (m^2)
m_f = Fuel load (wood equivalent) (kg)
A_o = Area of ventilation opening (m^2)
k_3 = 1.3–1.5, depending on the stick spacing in the cribs used as fuel (min m^2/kg)
A = Surface area of interior of enclosure (walls, floor, ceiling, and openings) (m^2)

In this correlation, A_{floor} was not included in the evaluation of solid surfaces because the floors were very well insulated.

In all experiments, the openings were the full compartment height.

The values of t-equivalent were found to be independent of the scale and height of ventilation openings.

Law then analyzed temperature–time data from a number of fires in larger brick and concrete compartments (approximately 3 m high) [115, 116] with fuels consisting of wood cribs, furniture, and liquid fuels and developed

$$t_e = k_4 m_f / (A_o [A - A_o])^{1/2} \qquad (C.5)$$

where k_4 is 1.0. This was due to the little effect fuel arrangement appeared to have in these larger scale tests.

In this correlation, the floor area was included in the evaluation of solid surfaces to which heat is lost. The larger scale data also showed no significant effect of ventilation opening height on t_e.

Law concluded this equation (C.5) was most suitable for engineering purposes for protected steel columns and went on to demonstrate that it gave good results for reinforced concrete also. She discovered that it overestimates the time prediction for tightly baled paper and cloth.

Pettersson

In 1976, Pettersson [117] adopted Law's approach to t-equivalent but, instead of the experimental curves on which her work was based, used the family of calculated temperature–time curves for particular compartments as derived by Magnusson and Thelandersson [118].

When the fuel load is expressed in mass (kg) of wood instead of "effective calorific value" (MJ), Pettersson's expression for t-equivalent is as follows:

$$t_e = 1.21 m_f / \left(A_o \sqrt{H_o} A \right)^{1/2} \qquad (C.6)$$

where:

H_o = Height of vertical opening (m)
A = Total area of internal envelope (walls, floor, ceiling, and openings). (Note that in his original heat balance work, he excluded A_o but for an unstated reason does not in his final equations presented in his design guide.)

This equation includes $\sqrt{H_o}$ because of the input parameters in the method for calculating the temperature–time curves on which this equation is based.

Equation C.6 can be modified to take into account the thermal properties of the compartment enclosure by applying the factor Ay to each input parameter.

This yields:

$$t_e = 1.21 k_f^{1/2} m_f'' A_{\text{floor}} / \left(A_o \sqrt{H_o} A \right)^{1/2} \qquad (C.7)$$

where k_f = factor applied to input parameters to take account of the thermal properties $k\rho c$ of the compartment enclosure expressed as a proportion of the $k\rho c$ for Pettersson's "standard" compartment. This is the compartment, defined in the Swedish Building Regulations in 1967, where the surrounding structure has the thermal properties of an average of concrete, brick, and lightweight concrete with a thickness of 20 cm. (Note also that the fire is ventilation-controlled and with a cooling phase of 10 °C/min.)

Normalized Heat Load Concept

In 1983, Harmathy and Mehaffey [119] developed the "normalized heat load" concept. The total heat penetrating the compartment boundaries is calculated taking into account $A\sqrt{H}$ and the proportion of heat evolution in the compartment, χ. When no unburnt gases emerge from the compartment, $\chi = 1$.

Based on the results of many experiments and tests using the Division of Building Research/ National Research Council of Canada floor test furnace [119], they derived the following relationship for t-equivalent: for t-equivalent:

$$t_e = 6.6 + 9.6 \times 10^{-4} H_N + 7.8 \times 10^{-9} H_N^2 \ (\text{s}) \tag{C.8}$$

for

$$0 < H_N < 9 \times 10^4$$

where:

$$H_N = 10^6(11.0\chi + 1.6)m_f'' A_{\text{floor}} / \left(A[k\rho c]^{1/2}\right.$$
$$\left. +1810\left\{A_o\sqrt{H_o}m_f'' A_{\text{floor}}\right\}^{1/2}\right) \left(\text{s}^{1/2}\,\text{K}\right) \tag{C.9}$$

$\chi = 0.41\left(H^3/A_o\sqrt{H_o}\right)^{1/2}$ or 1, whichever is less
H = Compartment height (m)
k = Thermal conductivity (kW/m K)
ρ = Density (kg/m^3)
c = Specific heat (kJ/kg K)
t_e from Eq. C.9 is then given approximately by:

$$t_e = 0.0016 H_N \tag{C.10}$$

for

$$H_N \leq 9 \times 10^4$$

Eurocode Time-Equivalent Method

The Eurocode [120] defines t-equivalent as described in the German standard DIN 18230, version 94, method [121]. The derivation of this formula has never been published, but it is understood to have come from an empirical analysis of calculated steel temperatures in a large number of simulated fires computed by the German program "Multi Room Fire Code." [121] Though this reference refers to an earlier published version of the Eurocode, the basic formulations remain, and therefore this origin is believed to still apply.

This method is dependent on ceiling height of the compartment but not the opening height. The fuel type assumed in the original work is unknown, though it is widely believed to be cellulosic.

The t-equivalent is defined in the Eurocode as:

$$t_{e,d} = q_{f,d} k_b w_f k_c \qquad (C.11)$$

where:

$q_{f,d}$ = Fuel load density related to the floor area (MJ/m^2), which can be calculated according to

$$q_{f,d} = q_{f,k} m \delta_{q1} \delta_{q2} \delta_n \qquad (C.12)$$

where:

$q_{f,k}$ = Fuel load density determined from a fuel load classification of occupancies (see Table C.1)
m = Combustion factor, which for cellulosic materials is defined as 0.8
$\delta_{q,1}$ = Safety factor taking account of the risk of a fire starting due to the size of compartment (see Table C.2)

Table C.1 Fuel load density determined from a fuel load classification of occupancies [120]

Occupancy	Average	80% Fractile
Dwelling	780	948
Hospital (room)	230	280
Hotel (room)	310	377
Library	1500	1824
Office	420	511
Classroom of a school	285	347
Shopping center	600	730
Theater (cinema)	300	365
Transport (public space)	100	122

Gumbel distribution is assumed for the 80% fractile

Table C.2 Safety factor taking account of the risk of a fire starting due to the size of compartment [120]

Danger of fire starting (δ_{q2})	Examples of occupancies
0.78	Art gallery, swimming pool
1.00	Offices, hotel, residential
1.22	Manufacturing for machinery and engines
1.44	Chemical lab, panting workshop
1.66	Manufacturing of fireworks or paints

Table C.3 Safety factor taking account of the risk of a fire starting due to the type of occupancy [120]

Compartment floor area A_f (m^2)	Danger of fire activation ($\delta_{q\,1}$)
25	1.1
250	1.5
2500	1.9
5000	2.0
10,000	2.13

$\delta_{q,2}$ = Safety factor taking account of the risk of a fire starting due to the type of occupancy (see Table C.3)

δ_n = Factor taking account of the different active fire-fighting measures such as sprinklers, detection, firefighters, etc.) (see Table C.4)

k_b is a conversion factor = 0.07 (min m^2/MJ) when no detailed assessment of the thermal properties of the boundary is pursued and when q_d is given in MJ/m^2.

Otherwise, k_b may be related to the thermal property $b = \sqrt{k\rho c}$ in accordance with Table C.5:

w_f is calculated as:

$$w_f = \left(\frac{6}{H}\right)^{0.3} \left[0.62 + \frac{90(0.4 - \alpha_v)^4}{1 + b_v \alpha_h}\right] \geq 0.5 \tag{C.13}$$

where:

$a_v = A_O/A_{\text{floor}}$ = Area of vertical openings A_O in the façade related to the floor area of the compartment where the limit $0.025 < av < 0.25$ should be observed

$\alpha_h = A_h/A_{\text{floor}}$ = Area of horizontal opening in the roof related to the floor area of the compartment

$b_v = 12.5(1+10\alpha_v - \alpha_v^2) \geq 10$

H = Height of the compartment (m)

For small fire compartments (defined in the Eurocode as $A_{\text{floor}} < 100$ m^2) without openings in the roof, the factor w_f may also be calculated as:

Table C.4 A factor taking account of the different active fire-fighting measures (sprinklers detection, fire fighters, etc.) [120]

δ_{ni} Function of active firefighting measures											
Automatic fire suppression				Automatic fire detection			Manual fire suppression				
Auto water extinguishing system	Independent water supplies (δ_{q2})			Auto fire detection and alarm		Auto transmission to fire brigade	Work fire brigade	Off-site fire brigade	Safe access routes	Fire-fighting devices	Smoke exhaust system
(δ_{n1})	0	1	2	By heat (δ_{n3})	By smoke (δ_{n4})	(δ_{n5})	(δ_{n6})	(δ_{n7})	(δ_{n8})	(δ_{n9})	(δ_{n10})
0.61	1	0.87	0.7	0.87 or 0.73		0.87	0.61 or 0.78		0.9/1/1.5	1/1.5	1/1.5

Note: According to the Eurocode, for "normal fire-fighting measures" such as safe access routes, firefighting devices, and smoke exhaust systems in staircases, the factors should be taken as 1.0, and if these measures have not been foreseen but provided, then the values can be taken as 1.5

Table C.5 Relationship between $k\&$ and the thermal property b

$b = \sqrt{k\rho c}$ J/m^2 s$^{1/2}$ K	K_b min m^2 /MJ
$b > 2500$	0.04
$720 \leq b <, 2500$	0.055
$b < 720$	0.07

Table C.6 Values for k_b recommended by the New Zealand Fire Engineering Design Guide

$\sqrt{k\rho c}$ (J/m^{2}Ks$^{1/2}$)	Construction materials	k_b value
400	Very light insulating materials	0.10
700	Plasterboard ceiling and walls, timber floor	0.09
1100	Lightweight concrete ceiling and floor, plasterboard wails	0.09
1700	Normal concrete ceiling and floor, plasterboard walls	0.065
2500	Thin sheet steel roof	0.045

$$w_f = \left(\frac{A_o\sqrt{H_o}}{A}\right)^{-\frac{1}{2}} \frac{A_{\text{floor}}}{A} \tag{C.14}$$

where:

$0.02 \leq \dfrac{A_o\sqrt{H_o}}{A} \leq 0.20$ with the default value $k_b = 0.07$, and assuming 18 MJ/kg for wood, Eq. C.14 becomes the same as Eq. C.7.

$Kc=$ A correction factor that is a function of the material composing structural cross-sections and is defined as $13.7\frac{A_o\sqrt{H_o}}{A}$ for unprotected steel. Reinforced concrete and protected concrete remain as 1.

The basis of this method is that it should be verified that $t_{e,d} < t_{fi,d}$ where $t_{fi,d}$ is the design value of the standard fire resistance of the members, assessed according to the relevant parts of the Eurocode.

This method could therefore be used for other defined periods of fire resistance such as in US codes.

New Zealand Code

The New Zealand Fire Engineering Design Guide [122] gives the same empirical expression for equivalent fire severity t_e (min) as the Eurocode.

The upper and lower kb values have been increased by a factor of 1.3 compared to the Eurocode due to what it declares inherent uncertainties in the Eurocode formula, the use of fuels other than wood, structures other than steel, and deep compartment effects. The values for kb recommended by the New Zealand *Fire Engineering Design Guide* are shown in Table C.6. If the properties of the linings are not known, a value of $k_b = 0.09$ is suggested. This formula is based on cellulosic-type fuels.

The ventilation factor limits of use are retained, though the small compartment formula in the Eurocode does not form part of the New Zealand guidance.

Comparisons

Time-equivalent methods are an improvement on the grading method in building codes worldwide, which is based on the standard fire temperature–time relationship (such as ASTM E119, BS 476 Part 20, or ISO834). This is because they attempt to account for compartment geometry, ventilation openings, fuel load density, and compartment boundary materials in addition to fuel load density, the key factors that affect full-scale fire development. However, the temperatures calculated on these principles are then related back to the standard temperature–time relationship. It is also important to note that they are based on specific compartment test data rather than generalized heat balance solutions. Time-equivalent methods are therefore unlike natural temperature–time relationships, which represent a real temperature–time relationship and are used as such, independent of the standard fire resistance test formulation.

Drysdale [14] describes a comparison Harmathy carried out where the Ingberg, Law, Pettersson, and Harmathy equations for t_e were compared. Drysdale rejects Ingberg's method since radiative heat flux varies with T^4, which makes simple scaling impossible when heat transfer is dominated by radiation. He concluded that Law and Harmathy provided more conservative solutions than the others. Note that Ingberg's method ignores ventilation, unlike the other methods presented here.

Law compared results using the time-equivalent relationships by Ingberg, Kawagoe, Law, Pettersson, Harmathy, and Mehaffey, plus the 1993 Eurocode formula with experimental data from post-flashover fires in full-scale compartments [115]. These consisted of small insulated compartments, 30 m^2 area, 2.5–3 m high, with brick or concrete enclosures [113], and larger, deeper rooms 128 m^2 in area (depth to width ratio 4:1) [123]. Law concluded Law, Pettersson, Harmathy, and Mehaffey were the most promising methods (Fig. C.2).

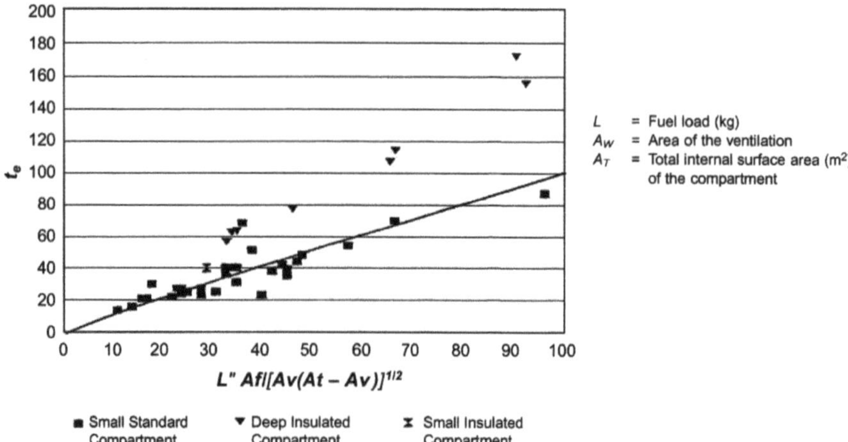

Fig. C.2 Law's correlation between fire resistance requirements (t_f) and $LI(A_WA_T)^{1/2}$ [115]

Limitations and Assumptions

The Deep Compartment Effect

Law examined deep compartments further since all her derived time-equivalent formulae gave odd results when deep compartments were studied. In deep compartments, temperatures and local burning rates are not uniform but rather progress from the opening toward the back of the enclosure as fuel is depleted [103]. Law also discovered that the 1993 Eurocode t-equivalent method gives poor correlation for both small and deep compartments [115].

She concluded that the depth of the compartment has an effect on time-equivalent over and above what can be allowed for by increase in insulation and in internal surface area A. Thomas and Heselden [114] had already shown that the ventilation-controlled rate of burning is affected by the compartment depth to width ratio.

Recent research on this phenomenon has also resulted in the New Zealand code's recommending factors of safety that have been increased by 30% to account for this effect in its time-equivalent formula.

The Eurocode

The Eurocode formulae do not reference the source of the equation derivations, particularly the ventilation factor needed in the time-equivalent calculation, as well as the correction factor to take account of cross-section material types, plus the other factors of safety recommended for application to the calculated time-equivalent value.

k_c is defined for unprotected steel as 13.7 times the opening factor. Since $0.02 < \dfrac{A_o \sqrt{H_o}}{A} < 0.02$, this gives $0.27 < k_c < 2.7$ for unprotected steel.

For small compartments ($A_{\text{floor}} < 100$ m^2) $w_f = \left(\dfrac{A_o\sqrt{H_o}}{A}\right)^{1/2} \dfrac{A_{\text{floor}}}{A}$, which can be written for unprotected steel in small compartments:

$$T_{ed} = 13.7 q_{fd} \times k_b \times \left(\frac{A_o\sqrt{H_o}}{A}\right)^{1/2} \frac{A_{\text{floor}}}{A}$$

(fuel load given as per unit floor area), Pettersson calculated T_e (h) for unprotected steel for values of $\dfrac{A_o\sqrt{H_o}}{A}$ ranging from 0.2 to 0.12 m$^{1/2}$.

Pettersson's time-equivalent formula is compared with the Eurocode formula for resultant emissivity, ε of 0.5 for unprotected steel as follows.

Pettersson

Section factor $= 50\ m^{-1}$

$Q_{t,d}$	$T_e(\text{h})\ \frac{A_o\sqrt{H_o}}{A}$	$= 0.02$	0.04	0.08	0.12
42		0.29	0.25	0.21	0.16
84		0.42	0.43	0.36	0.29
126		0.51	0.58	0.525	0.41

Section factor $= 150\ m^{-1}$

$Q_{t,d}$	$T_e(\text{h})\ \frac{A_o\sqrt{H_o}}{A}$	$= 0.02$	0.04	0.08	0.12
42		0.21	0.23	0.19	0.16
84		0.27	0.37	0.38	0.30
126		0.30			

It can be seen that, while the values of $\underline{T_e}$ tend to increase with increasing fuel load, they tend to decrease with the ventilation factor. They are not independent of the section factor.

Yet when the Eurocode formula with $k_b = 0.055$ for Pettersson's compartment type A, for all sections, is used to do a similar check, the following data is produced:

Eurocode

$Q_{t,d}$	$T_e(\text{h})\ \frac{A_o\sqrt{H_o}}{A}$	$= 0.02$	0.04	0.08	0.12
42		0.07	0.11	0.15	0.18
84		0.15	0.21	0.30	0.37
126		0.22	0.32	0.45	0.55

The trends in the Eurocode are different, and no explanation as to why has been provided.

For unprotected steel, if the steel is assumed to be at a uniform temperature, the post-flashover fire temperatures are also perfectly stirred, and the temperature of the exposed surface is the same as the fire temperatures; this implies the steel temperature is always the same as the fire temperature. Therefore, section factors of $100\ m^{-1}$ or more would not be expected to survive a post-flashover fire but a localized fire.

Until suitable justification of such difference is made, it seems the original work of Pettersson or Law is best suited to this type of time-equivalent calculation.

Appendix D: Examples

Example 1

A room 5 m wide, 4 m deep, and 2.5 m high has one vent that is 2 m high and 3 m wide. The fuel load is 10 lb/ft^2. The enclosure is made of gypsum plaster on metal with the following properties:

$$k = 0.47 \ \text{W/m}^\circ\text{C}$$
$$\rho = 1440 \ \text{kg/m3}$$
$$c = 0.84 \ \text{kJ/kg}^\circ\text{C}$$

Find the maximum temperature of the fire and its burning duration.

Law's method is recommended for all roughly cubic compartments and for long, narrow compartments where $\frac{A}{A_o\sqrt{H_o}}$ is approximately less than or equal to 18 m$^{-1/2}$. Since this room is roughly cubic, Law's method is applicable.

$A = 2(5*4) + 2(5*25) + 2(4*3.5) = 85 \ \text{m}^2$
$A_o = 2*3 - 6 \ \text{m}$
$H_o = 2 \ \text{m}$

$$\frac{A}{A_o\sqrt{H_o}} = \frac{85}{6*\sqrt{2}} = 10.02 \ \text{m}^{-1/2}$$

Maximum Temperature

$$T_{(\text{max})} = 6000 \left(\frac{1 - e^{-0.1\frac{A}{A_o\sqrt{H_o}}}}{\sqrt{\frac{A}{A_o\sqrt{H_o}}}} \right)$$

Thus,

$$T_{(\text{max})} = 6000 \left(\frac{1 - e^{-0.1*10.02}}{\sqrt{10.02}} \right) = 1199.56^\circ\text{C} \quad (\text{ans.})$$

Burning Duration

$$\tau = \frac{m_f}{\dot{m}_f}$$

where:

$m_f = m''_f \times A_{\text{floor}}$
$m''_f = 10 \text{ lb/ft}^2$ or 49 kg/m^2
$A_{\text{floor}} = 20 \text{ m}^2$
$m_f = 49 \times 20 = 980 \text{ kg}$

and

$$\dot{m}_f = 0.18 A_o \sqrt{H_o}\left(\frac{W}{D}\right)\left(1 - e^{-0.036\frac{A}{A_o\sqrt{H_o}}}\right)$$

where W is width of the room and D is depth of the room.

$$\dot{m}_f = 0.18^*6^*\sqrt{2}^*\left(\frac{5}{4}\right)\left(1 - e^{-0.036 \times 10}\right) = 0.578 \text{kg}\!/_{\!s}$$

This equation is valid for

$$\frac{\dot{m}_f}{A_o\sqrt{H_o}}\left(\frac{D}{W}\right)^{1/2} < 60$$

and in this case

$$\frac{0.578}{6\sqrt{2}}\left(\frac{4}{5}\right)^{1/2} = 0.061 < 60$$

To ensure that predictions are sufficiently conservative using Law's method, the predicted burning rate should be reduced by a factor of 1.4.
The adjusted burning rate is then

$$\dot{m}_f = \frac{\dot{m}_f}{1.4} = \frac{0.578}{1.4} = 0.413 \text{kg}\!/_{\!s}$$

The burning duration can be found by

$$\tau = \frac{m_f}{\dot{m}_f} = \frac{980}{0.413} = 2400 \text{ s} = 0.66 \text{ h}$$

Example 2

A room 7 m wide, 28 m deep, and 4 m high has one vent that is 4 m high and 3.5 m wide in one of the small end walls. The fuel load is 35 kg/m^2. The enclosure is made of brick with the following properties:

$$k = 0.69 \text{ W/m}\,^\circ\text{C}$$
$$\rho = 1600 \text{ kg/m3}$$
$$c = 0.84 \text{ kJ/kg}\,^\circ\text{C}$$

Find the burning duration, and plot the temperature–time curve.

For long, narrow spaces in which the value of $\frac{A}{A_o\sqrt{H_o}}$ is in the range of 45–85 m$^{-1/2}$, Magnusson and Thelandersson's method is recommended.

$$A = 2(7^*4) + 2(7^*28) + 2(4^*28) = 672 \text{ m}^2$$
$$A_o = 1.9^*3.7 = 7 \text{ m}$$
$$H_o = 3.7 \text{ m}$$

In this case,

$$\frac{A}{A_o\sqrt{H_o}} = \frac{672}{7\sqrt{3.7}} = 50 \text{ m}^{-1/2},$$

and Magnusson and Thelandersson's method is used.

The first step is to decide which of the seven models in Magnusson and Thelandersson's method is applicable to the problem.

Type A enclosed spaces consist of a material, 20 cm in thickness, whose thermal properties are characterized by the following average values. These characteristics normally belong to concrete, brick, and lightweight concrete.

Thermal conductivity: $k = 0.7$ kcal/m-h-°C

Product of the specific heat and the density, $c * \rho = 400$ kcal/m^{3o}C

In this case, $k = 0.5937$ lccal/mh°C and $c* \rho = 321.22$ kcal/m^{3o}C. Thus, Type A enclosed space is used.

The next step is to calculate the burning duration, τ.

$$\tau = \frac{qA}{1500A_o\sqrt{H_o}}$$

q, the fire load per bounding surface area, is calculated using the fuel load and the heat of combustion, ΔH_c. The heat of combustion of cellulosic materials, 15 MJ/kg, is used for this example.

$$q = m_f'' \times \frac{A_{\text{floor}}}{A} \times \Delta H_c$$
$$q = 35 \text{ kg/m}^{2*}(7^*28)/672^*15 \text{ MJ/kg}$$
$$= 153 \text{ MJ/m}^2 = 37 \text{ Mcal/m}^2$$

Therefore,

$$\tau = \frac{37 \times 672}{1500^*7^*\sqrt{3.7}} = 1.2 \text{ h} \quad \text{(ans.)}$$

The type of enclosure, opening factor, and burning duration can be used to reference Magnusson and Thelandersson's tables. The tables give the temperature at 0.05-, 0.10-, and 0.20-h intervals up to 6.00 h for various burning durations. The temperatures for a burning duration of 1.5 h and an opening factor of 0.02 m$^{1/2}$ were used to create the temperature–time curve in Fig. D.1.

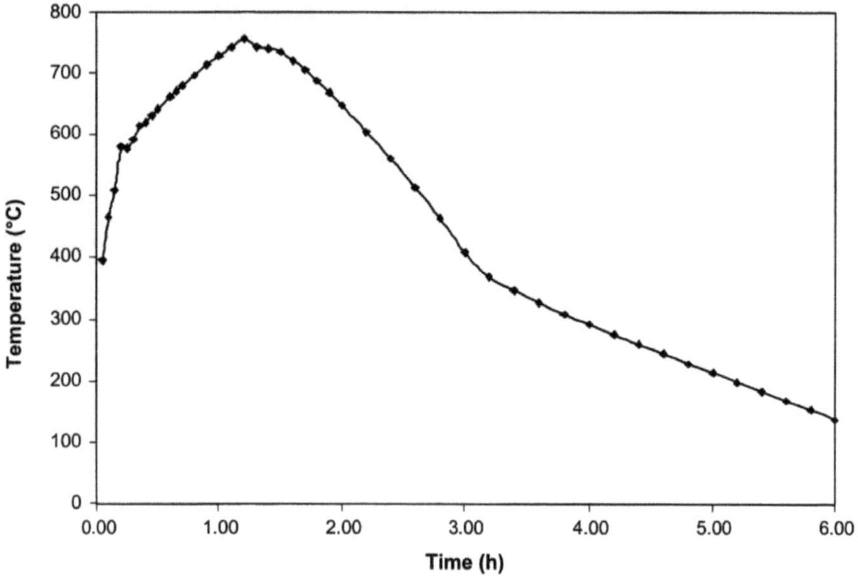

Fig. D.1 Temperature–time curve for burning duration of 1.5 h and opening factor of 0.02 m$^{1/2}$

Glossary

Nomenclature Used in the Enclosure Fires Section

A	Surface area of the interior of the enclosure (m^2)
A_f	Surface area of fuel (m^2)
A_{floor}	Surface area of floor (m^2)
A_o	Area of ventilation opening (m^2)
b	Stick width (m)
b_p	Factor $(-)$
C	Wood constant (g/m$^{1.5}$-s)
c	Specific heat of enclosure lining (J/kg-K)
c_p	Specific heat of air (J/kg-K)
D	Depth of the compartment (m)
F	View factor $(-)$ or opening factor (m$^{1/2}$)
G	Gravitational constant (9.81 m/s^2)
h	Equivalent conductance (W/m^2-K)
h_c	Convection coefficient (W/m^2-K)
h_k	Conduction coefficient (W/m^2-K)
h_r	Radiation coefficient (W/m^2-IC)
H	Height of compartment (m)
H_o	Height of ventilation opening (m)
k	Thermal conductivity of enclosure lining (W/m-K)
k_o	Coefficient $(-)$
L	Latent heat of vaporization (lcj/g)
m_f	Mass of fuel (kg)
$\dot{m}_f$	Mass burning rate of fuel (kg/s)
m''_f	Mass of fuel per unit area (lcg/m^2)
$\dot{m}''_f$	Mass burning rate of fuel per limit area (kg/m^2-s)
$\dot{m}''_{f,o}$	Free burning mass loss rate of fuel per unit area (kg/m^2-s)
$\dot{m}''_{f,\infty}$	Asymptotic mass loss rate of fuel per unit area (kg/m^2-s)

Fire Exposures to Structural Elements, The Society of Fire Protection Engineers Series,
https://doi.org/10.1007/978-3-031-64025-4

$\dot{m}_0$	Mass flow rate of air (kg/s)
p	Pressure (Pa)
q	Fuel load density (Mcal/m^2)
$\dot{q}$	Heat loss rate (kW)
$\dot{q}_w$	Heat loss through walls (kW)
$\dot{q}''_{\text{fire}}$	Heat flux from fire (kW/m^2)
$\overline{q}_E$	Effective heat flux (W/m^2)
$\dot{Q}$	Heat release rate (kW)
$\dot{Q}_B$	Rate of the heat energy stored in the gas volume
$\dot{Q}_L$	Rate of heat energy withdrawn from the enclosed space due to airflow
$\dot{Q}_R$	Rate of heat energy withdrawn from the enclosed space by radiation
$\dot{Q}_W$	Rate of heat energy withdrawn from enclosed space through the wall, floor, or ceiling
Q^*	Dimensionless heat release rate ($-$)
Q_f^*	Dimensionless radiation rate to fuel ($-$)
Q_r^*	Dimensionless radiation loss rate ($-$)
Q_w^*	Dimensionless heat loss rate to walls ($-$)
R	Universal gas constant (831 J/kMol-K)
s	Stoichiometric air to fuel ratio ($-$)
t	Time (units as stated)
t_m	Time corresponding to maximum temperature (units as stated)
T	Temperature in compartment (units as stated)
T_b	Fuel boiling point (units as stated)
T_f	Flame temperature (units as stated)
T_{gm}	Maximum temperature (units as stated)
T_{gmcr}	Maximum temperature in the critical region (units as stated)
T_o	Ambient temperature (units as stated)
T_w	Wall temperature (units as stated)
V	Volume (m^3)
W	Width of wall containing ventilation opening (m)
Y_{O_2}	Mass fraction of O_2 ($-$)

Greek

β	Factor ($-$)
Γ	Scaling factor
ϕ	Equivalence ratio ($-$)
δ	Thickness (m) or shape factor ($-$)
ΔH_p	Heat of vaporization of liquid (kJ/kg)
ΔH_c	Heat of combustion (MJ/kg)
ΔH_{air}	Heat of combustion per unit mass of air (MJ/kg)
ε	Gas emissivity ($-$)
ε_w	Wall emissivity ($-$)

κ	Absorption coefficient (m^{-1}) or factor ($-$)
θ_1-θ_5	Variable ($-$)
η	Factor ($-$)
η_{cr}	Factor ($-$)
ρ	Density of enclosure lining (kg/m^3)
ρ_0	Density of air (kg/m^3)
σ	Stefan-Boltzmann Constant [5.67×10^{11} kW/(m^2 K^4)]
τ	Duration of fully developed fire (units as stated)
ν	Kinematic viscosity (m^2/s)
ς	Factor ($-$)
Ψ	Factor (kg/m^2)

Nomenclature Used in the Plumes Section

A	Surface area of noncircular fuel package (m^2)
b_u	Plume width (m)
b_t	Thermal plume width (m)
C_p	Specific heat capacity of air at 300 K [1.0 kJ/(kg K)]
D	Length of single side of square burner, diameter (m)
g	Acceleration of gravity (9.81 m/s^2)
h	Convective heat transfer coefficient [kW/(m^2 K)]
H	Distance between base of fire and ceiling (m)
H_B	Distance between base of fire and lower flange of I-beam (m)
H_C	Distance between base of fire and upper flange of I-beam (m)
H_W	Distance between the base of fire and center of the web on I-beam (m)
h	Convective heat transfer coefficient [kW/(m K)]
L_B	Distance from fire centerline to flame tip along the lower flange of an I-beam (m)
L_C	Distance from fire centerline to flame tip along the upper flange of an I-beam (m)
L_H	Distance from fire centerline to flame tip length along the ceiling or upper flange of an I-beam (m)
L_W	Distance from fire centerline to flame tip length along the web center of an I-beam (m)
L_f	Average flame length or unconfined flame tip length (m)
$L_{f,\text{tip}}$	Flame tip length (m)
$L_{f,\text{tip}B}$	Flame tip length along the lower flange of I-beam (m)
$L_{f,\text{tip}C}$	Flame tip length along the upper flange of I-beam (m)
$L_{f,\text{tip}W}$	Flame tip length along web center of an I-beam (m)
Q	Fire heat release rate (kW)
Q^*	Dimensionless parameter, $Q_D^* \dfrac{Q}{\rho_\infty C_p T_\infty \sqrt{g} D^{5/2}}$ being D length scale
r	Distance from comer or stagnation point to measurement location or radial distance for plume centerline (m)

$\dot{q}''$	Heat flux (kW/m^2)
T	Temperature (K)
$T_{m,c}$	Centerline plume temperature (K)
T_g	Room gas temperature (K)
T_s	Material surface temperature (K)
T_∞	Ambient temperature (300 K)
U	Plume velocity (m/s)
$U_{m,c}$	Centerline plume velocity (m/s)
w	Dimensionless distance along ceiling or I-beam, $w = (r + H + z')/(L_H + H + z')$
x	Horizontal distance from the comer or fire centerline or width distance into the material thickness (m)
y	Horizontal distance from comer (m)
z	Vertical distance above the base of the fire (m)
z'	Virtual source origin correction in tests with fires impinging on ceilings and I-beams (m)
z_o	Virtual source origin correction for plumes (m)

Greek

α	Absorptivity (- -)
χ_r	Radiative fraction (- -)
ε	Emissivity (- -)
ρ_∞	Ambient density of air (1.2 kg/m^3)
π	Constant (3.14159)
σ	Stefan-Boltzmann constant [5.67×10^{-11} $kW/(m^2\ K^4)$]

Subscripts

cl	Centerline
conv	Convective
D	Defined using D as length scale
f	Flame
hfg	Heat flux gauge
H	Defined using H as the length scale
H_B	Defined using H_B as the length scale
H_C	Defined using H_C as the length scale
inc	Incident
m	Measured
max	Maximum level
net	Net
peak	Peak

rad	Radiative
rr	Reradiated
s	Material surface
w	Centerline of web

References

1. Klem, T.J., Fire Investigation Report, One Meridian Plaza, Philadelphia, Pennsylvania, Three Fire Fighter Fatalities, February 23, 1991, Quincy, Mass., National Fire Protection Association, 1991.
2. "'Smoke Caused Most Damage' in Broadgate Fire," *New Builder* **90**:6 (July 11, 1991).
3. *SFPE Engineering Guide to Performance-Based Fire Protection Analysis and Design of Buildings, Quincy, Mass., National Fire Protection Association, 2000.*
4. *International Code Council Performance Code,* Falls Church, Va., International Code Council, 2003.
5. NFPA 5000, *Building Construction and Safety Code,* Quincy, Mass., National Fire Protection Association, 2003.
6. Eurocode – Basis of Design and Actions on Structures, Part 2.2: Actions on Structures – Actions on Structures Exposed to Fire, ENV 1991-2-2, CEN, 1995.
7. Babrauslcas, V., COMPF: A Program for Calculating Post-flashover Fire Temperatures, UCB FRG *75-2*, Berkeley, University of California, Fire Research Group, 1975.
8. McGrattan, K., et al., "Fire Dynamics Simulator (Version 3) – Technical Reference Guide," NISTIR 6783, Gaithersburg, Md., National Institute of Standards and Technology, 2002.
9. McGrattan, K., et al., "Fire Dynamics Simulator (Version 3) – User's Guide," NISTIR 6784, Gaithersburg, Md., National Institute of Standards and Technology, 2002.
10. Buchanan, A., *Structural Design for Fire Safety,* [London], John Wiley and Sons, 2001.
11. "Rational Fire Safety Engineering Approach to Fire Resistance in Buildings," CIB W14 Report 269, Rotterdam, International Council for Research and Innovation in Building and Construction, 2001.
12. Karlsson, B., and J. Quintiere, *Enclosure Fire Dynamics,* Boca Raton, Fla., CRC Press, 2000.
13. Thomas, I., and I. Bennets, "Fires in Enclosures with Single Ventilation Openings – Comparison of Long and Wide Enclosures," *Fire Safety Science – Proceedings of the Sixth International Symposium,* London, International Association for Fire Safety Science, 1999.
14. Drysdale, D., *Introduction to Fire Dynamics,* Chichester, U.K., John Wiley & Sons, 1999.
15. Karlsson, B., and J. Quintiere, *Enclosure Fire Dynamics,* Boca Raton, Fla., CRC Press, 2000.
16. Thomas, P., and A. Heselden, "Fully Developed Fires in Single Compartments," Co-operative Research Programme of the Conseil International du Batiment, CIB Report No. 20, *Fire Research Note No. 923,* Borehamwood, England, Fire Research Station, 1972.
17. Tewarson, A., "Generation of Heat mid Chemical Compounds in Fires," *The SFPE Handbook of Fire Protection Engineering,* Quincy, Mass., National Fire Protection Association, 2002.
18. Gross, D., "Experiments on the Burning of Cross Piles of Wood," *Jour, of Research of the Nat. Bureau of Standards-C, Engineering and Instrumentation 66C:2* (April–June 1962) 99.

© Society for Fire Protection Engineers 2025

187

Fire Exposures to Structural Elements, The Society of Fire Protection Engineers Series,
https://doi.org/10.1007/978-3-031-64025-4

19. Heskestad, G., "Modeling of Enclosure Fires," *14th Symposium (Int.) on Combustion,* Pittsburgh, Pa., The Combustion Institute, 1973, p. 1021.

20. Harmathy, T., "A New Look at Compartment Fires, Part I," *Fire Technology* **8**:2 (1972) 196–217.

21. Babrauslcas, V., "Heat Release Rates," *The SFPE Handbook of Fire Protection Engineering,* Quincy, Mass., National Fire Protection Association, 2002.

22. McCaffrey, B., J. Quintiere, and M. Harkelroad, "Estimating Room Fire Temperatures and the Likelihood of Flashover Using Fire Test Data Correlations," *Fire Technology* **17**:2 (1981) 98–119.

23. Tanaka, T., M. Sato, and T. Wakamatsu, "Simple Formula for Ventilation Controlled Fire Temperatures," 13th Meeting of the U.S./Japan Government Cooperative Program on Natural Resources Panel on Fire Research and Safety, March 13–20, 1996, NISTIR 6030, Gaithersburg, Md., National Institute of Standards and Technology, 1997.

24. Quintiere, J., "Fire Behavior in Building Compartments," *Proc. of the Combustion Institute,* Vol. 29, Pittsburgh, The Combustion Institute, 2002.

25. Wickstrom, U., "Application of the Standard Fire Curve for Expressing Natural Fires for Design Purposes," *Fire Safety: Science and Engineering,* ASTM STP 882, T.Z. Harmathy, ed., Philadelphia, American Society for Testing and Materials, 1985, pp. 145–159.

26. Franssen, J., "Improvement of the Parametric Fire of Eurocode 1 Based on Experimental Test Results," *Fire Safety Science – Proceedings of the Sixth International Symposium,* London, International Association for Fire Safety Science, 2000.

27. Dinenno, P., ed., *The SFPE Handbook of Fire Protection Engineering,* Appendix B, Quincy, Mass., National Fire Protection Association, 2002.

28. Thomas, P., et al., "Design Guide – Structural Fire Safety," *Fire Safety Journal* **10**:2 (March 1986) 75–137.

29. Kumar, S., and V. Rao, "Fire Loads in Office Buildings," *Journal of Structural Engineering* (March 1997) 365–368.

30. Narayanan, P., "Fire Severities for Structural Engineering Design," Study Report No. 67, Wellington, New Zealand, BRANZ, 1996.

31. Caro, T., and J. Milke, "A Survey of Fuel Loads in Contemporary Office Buildings," NIST-GCR-96-697, Gaithersburg, Md., National Institute of Standards and Technology, 1996.

32. Dinenno, P., ed., *The SFPE Handbook of Fire Protection Engineering,* Appendix C, Quincy, Mass., National Fire Protection Association, 2002.

33. Tewarson, A., "Generation of Heat and Chemical Compounds in Fires," *The SFPE Handbook of Fire Protection Engineering,* Quincy, Mass., National Fire Protection Association, 2002.

34. Drysdale, D., *An Introduction to Fire Dynamics.* 2nd ed., Chichester, U.K., Wiley & Sons, 1999.

35. Kawagoe, K., and T. Sekine, "Estimation of Fire Temperature–Time Curve in Rooms," Occasional Report No. 11, Tokyo, Building Research Institute, 1963.

36. Lie, T., "Fire Temperature–Time Relationships," *The SFPE Handbook of Fire Protection Engineering,* 2nd ed., Quincy, Mass., National Fire Protection Association, 1995.

37. Tanaka, T., et al., "Simple Formula for Ventilation Controlled Fire Temperatures," U.S./Japan Government Cooperative Program on Natural Resources (UJNR), Fire Research and Safety, 13th Joint Panel Meeting, K.A. Beall, ed., Gaithersburg, Md., [National Institute of Standards and Technology], 1996.

38. Magnusson, S.E., and S. Thelandersson, "Temperature–Time Curves of Complete Process of Fire Development," Civil Engineering and Building Construction Series No. 65, Stockholm, ACTA Polytechnica Scandinavica, 1970.

39. Harmathy, T., "A New Look at Compartment Fires, Part I," *Fire Technology* **8**:2 (1972) 196–217.

40. Harmathy, T., "A New Look at Compartment Fires, Part II," *Fire Technology* **8**:4 (1.972) 326–351.

41. Babrauskas, V., "Fire Endurance in Buildings," Ph. D. Dissertation, Berkeley, Univ. of California, 1976.

42. Babrauskas, V., and R.B. Williamson, "Post-Flashover Compartment Fires: Basis of a Theoretical Model," *Fire and Materials 2 (1978) 39–53.*

43. Babrauskas, V., and R.B. Williamson, "Post-Flashover Compartment Fires – Application of a Theoretical Model," *Fire and Materials* 3 (1979) 1–7.

44. Babrauskas, V., COMPF2 – A Program for Calculating Post-Flashover Fire Temperatures, Tech Note 991, Gaithersburg, Md., National Bureau of Standards, 1979.

45. Babrauskas, V., "A Closed-Form Approximation for Post-Flashover Compartment Fire Temperatures," *Fire Safety Journal 4 (1981) 63–73.*

46. Babrauskas, V., "Pleat Release Rates," *The SFPE Handbook of Fire Protection Engineering,* Quincy, Mass., National Fire Protection Association, 2002.

47. Ma, Zhongcheng, and Pentti Makelainen, "Parametric Temperature–Time Curves of Medium Compartment Fires for Structural Design," *Fire Safety Journal 34* (2000) 361–375.

48. Thomas, P., and A. Heselden, "Fully Developed Fires in Single Compartments," A Co-operative Research Program of the Conseil International du Bathnent, CEB Report No. 20, Fire Research Note No. 923, Borehamwood, England, Fire Research Station, 1972.

49. Law, M., "A Basis for the Design of Fire Protection of Building Structures," *The Structural Engineer* 61 **A**: 5 (January 1983).

50. Kawagoe, K., "Fire Behavior in Rooms," Report 27, Tokyo, Building Research Institute, Ministry of Construction, 1958.

51. Lie, T.T., "Characteristic Temperature Curves for Various Fire Severities," *Fire Technology 10* (1974) 315–326.

52. Lie, T.T., *Fire and Buildings,* London, Applied Science Publishers Ltd., 1972, pp. 8–22.

53. Lie, T.T., *Structural Fire Protection,* New York, American Society of Civil Engineers, 1992, pp. 137–158.

54. Lattimer, B.Y., "Heat Fluxes from Fires to Surfaces," *The SFPE Handbook of Fire Protection Engineering,* 3rd ed., Quincy, Mass., National Fire Protection Association, 2002.

55. Beyler, C., "Fire Plumes and Ceiling Jets," *Fire Safety Journal 11* (1986) 53–75.

56. Cox, G., and R. Chitty, "Some Source-Dependent Effects of Unbounded Fires," *Combustion and Flame* **60** (1985) 219–232.

57. Heskestad, G., "Virtual Origins of Fire Plumes," *Fire Safety Journal* **5** (1983) 109–114.

58. Heskestad, G., "Luminous Height of Turbulent Diffusion Flames," *Fire Safety Journal* **5**: 103–108.

59. Zulcosld, E., T. Kubota, and B. Cetegen, "Entrainment in Fire Plumes," *Fire Safety Journal* **3** (1981).

60. Quintiere, J.G., and B.S. Grove, *27th Symp. (Int.) on Combustion,* Comb. Inst. 2757, Pittsburgh, The Combustion Institute, 1998.

61. Smith, D.A., and G. Cox, *Combustion and Flame* **60** (1992) 219.

62. Kung, H.C., and P. Stavrianidis, *19th Symp. (Int) on Combustion,* Pittsburgh, The Combustion Institute, 1983.

63. Baum, H.R., and B.J. McCaffrey, *Fire Safety Science – Proceedings of the Second International Symposium,* New York, Hemisphere Pub. Corp., 1989, p. 129.

64. Koseki, H., *Fire Safety Sci. – Proceedings of the 6th Int. Symp.,* Int. Assoc, for Fire Safety Sci., 2000, p. 115.

65. McCaffrey, B., NBSIR 79-1910, Washington, D.C., National Bureau of Standards, 1979.

66. *Assessing Flame Radiation to External Targets from Pool Fires,* SFPE Engineering Guide, Bethesda, Md., Society of Fire Protection Engineers, 1999.

67. Beyler, C., "Fire Hazard Calculations for Large, Open Hydrocarbon Pool Fires," *The SFPE Handbook of Fire Protection Engineering,* 3rd ed., Quincy, Mass., National Fire Protection Association, 2002,

68. Baum, H.R., and B.J. McCaffrey, "Fire Induced Flow Field – Theory and Experiment," *Fire Safety Science - Proceedings of the Second International Symposium,* New York, Hemisphere Pub. Corp., 1989.

69. Babrauslcas, V., "Pleat Release Rates," *The SFPE Handbook of Fire Protection Engineering,* 3rd ed., Quincy, Mass., National Fire Protection Association, 2002,

70. Hoglander, IC., and B. Sundstrum, "Design Fires for Preflashover Fires – Characteristic Heat Release Rates of Building Contents," SP Report 1997:36, Boras, Sweden, Swedish National Testing and Research Institute, 1997.

71. Mudan, K.S., and P.A. Croce, "Fire Hazard Calculations for Large Open Hydrocarbon Fires," *The SFPE Handbook of Fire Protection Engineering,* 2nd ed., Quincy, Mass., National Fire Protection Association, 1995.

72. Tewarson, A., "Generation of Heat and Chemical Compounds in Fires," Section 3, Chapter 4, *The SFPE Handbook of Fire Protection Engineering,* 3rd ed., Quincy, Mass., National Fire Protection Association, 2002.

73. Evans, D., *Fire Technology* **20** (1984).

74. Cooper, L., *Proceedings of the 19th International Symposium on Combustion,* 1982.

75. Heskestad, G., "Fire Plumes, Flame Height, and Air Entrainment," Section 2, Chapter 1, *The SFPE Handbook of Fire Protection Engineering,* 3rd ed., Quincy, Mass., National Fire Protection Association, 2002.

76. Quintiere, J., and B. Grove, "Correlations for Fire Plumes," NIST-GCR-98-744, Gaithersburg, Md., National Institute of Standards and Technology, 1998.

77. Kokkala, M., "Heat Transfer to and Ignition of Ceiling by an Impinging Diffusion Flame," VTT Research Report 586, Finland, VTT, 1989.

78. Kokkala, M., "Experimental Study of Heat Transfer to Ceiling from an Impinging Diffusion Flame," *Fire Safety Scienc – Proceedings of the Third International Symposium,* International Association for Fire Safety Science, 1991.

79. You, H., and G. Faeth, "Ceiling Heat Transfer During Fire Plume and Fire Impingement," *Fire and Materials* **3**:3 (1979) 140–147.

80. You, H., and G. Faeth, "An Investigation of Fire Impingement on a Horizontal Ceiling," NBS-GCR-79-188, Gaithersburg, Md., National Bureau of Standards, 1979.

81. Gregory, J., R. Mata, and N. Keltner, "Thermal Measurements in a Series of Large Pool Fires," Sandia Report Number SAND85-0196, Albuquerque, N.M., Sandia National Laboratories, 1987.

82. Russell, L., and J. Canfield, "Experimental Measurements of Heat Transfer to a Cylinder Immersed in a Large Aviation Fuel Fire," *Journal of Heat Transfer* (August 1973).

83. Cowley, L., "Behaviour of Oil and Gas Fires in the Presence of Confinement and Obstacles," Miscellaneous Report TNMR.91.006, Chester, U.K., Shell Research Ltd., Thornton Research Center, Combustion and Fuels Department, 1991.

84. Back, G., et al., "Wall Incident Heat Flux Distributions Resulting from an Adjacent Fire," *Fire Safety Science – Proceedings of the Fourth International Symposium,* International Association for Fire Safety Science, 1994.

85. Gross, D., and J. Fang, "The Definition of a Low Intensity Fire," NBS Special Publication 361, Voi. 1: *Performance Concept in Buildings – Proceeding of the Joint RILEM-ASTM-CIB Symposium,* 1972.

86. Mizuno, T., and K. Kawagoe, "Burning Behaviour of Upholstered Chairs: Part 2. Burning Rate of Chairs in Fire Tests," *Fire Science and Technology* **5**:1 (1985) 69–78.

87. Quintiere, J., and T. Cleary, "Heat Flux from Flames to Vertical Surfaces," *ASME – Heat and Mass Transfer in Fire and Combustion Systems, HTD 223* (1992) 111–120.

88. Quintiere, J., and T. Cleary, "Heat Flux from Flames to Vertical Surfaces," *Fire Technology* **30**:2 (1994) 209–231.

89. Dillon, S.E., "Analysis of the ISO 9705 Room/Comer Test: Simulations, Correlations and Heat Flux Measurements," NIST-GCR-98–756, [Gaithersburg, Md.], National Institute of Standards and Technology, 1998.

90. Williamson, R., A. Revenaugh, and F. Mowrer, "Ignition Sources in Room Fire Tests and Some Implications for Flame Spread Evaluation," *Fire Safety Science – Proceedings of the Third International Symposium*, International Association for Fire Safety Science, 1991.

91. Lattimer, B., et al., "Development of a Model for Predicting Fire Growth in a Combustible Comer," NSWC Report, NSWCCD-TR-64-99/07, U.S. Navy, 1999.

92. Alpert, R., "Turbulent Ceiling Jet Induced by Large-Scale Fires," *Combustion Science and Technology 11* (1975).

93. Kokkala, M., "Characteristics of a Flame in an Open Comer of Walls," *Proceedings from INTERFLAM '93*, Interscience, 1993.

94. Hasemi, Y., et al, "Flame Length and Flame Heat Transfer Correlations in Comer-Wall and Comer-Wall-Ceiling Configurations," *Proceedings from INTERFLAM '96*, Interscience, 1996.

95. Ohlemiller, T, T. Cleary, and J. Shields, "Effect of Ignition Conditions on Upward Flame Spread on a Composite Material in a Corner Configuration," *Fire Safety Journal* **31** (1998).

96. Janssens, M., and H. Tran, "Modeling the Burner Source Used in ASTM Room Fire Test," *Journal of Fire Protection Engineering* **5**:2 (1993) 53–66.

97. Hasemi, Y, et al., "Fire Safety of Building Components Exposed to a Localized Fire – Scope and Experiments on Ceiling/Beam System Exposed to a Localized Fire," *Proceedings of ASIAFLAM*, 1995.

98. Myllymald, J., and M. Kokkala, "Thermal Exposure to a High Welded I-Beam Above a Pool Fire," *First International Workshop on Structures in Fires*, Copenhagen, 2000.

99. Alpert, R., "Convective Heat Transfer in the Impingement Region of a Buoyant Plume," *Transactions of ASME 109* (1987) 120–124.

100. You, H., "An Investigation of Fire-Plume Impingement on a Florizontal Ceiling 2 – Impingement and Ceiling-Jet Regions," *Fire and Materials* **9**:1 (1985) 46–56.

101. Cooper, L., "Heat Transfer from a Buoyant Plume to an Unconfhied Ceiling," *ASME Journal of Heat Transfer 104* (1982) 446–452.

102. Wakamatsu, T, et al., "Experimental Study on the Heating Mechanism of a Steel Beam Under Ceiling Exposed to a Localized Fire," *Proceedings from INTERFLAM '96*, Interscience, 1996.

103. Kirby, B., et al., "Natural Fires in Large-Scale Compartments – A British Steel Technical, Fire Research Station Collaborative Project," Rotterdam, U.K., British Steel, 1994.

104. http://www.zaintraders.com/thermosil.html.

105. ASTM E119-95a, "Standard Test Methods for Fire Tests of Building Construction and Materials," West Conshohocken, Pa., American Society for Testing and Materials, 1995.

106. BS 476, "Fire Tests on Building Materials and Structures," Part 20, "Method for Determination of tire Fire Resistance of Elements of Construction (General Principles)," British Standards Institution, 1987.

107. ISO 834, "Fire-Resistance Tests – Elements of Building Construction," Geneva, International Organization for Standardization, 1999.

108. "Competitive Steel Buildings Through the Natural Fire Safety Concept," Final Report No. 32, CEC Agreement 7210-SA/125, 126, 213, 214, 323, 423, 522,623, 839, 937, Profil ARBED Centre de Recherches, March 1999.

109. *Fire Protection Through Modern Building Codes*, 5th ed., Washington, American Iron and Steel Institute, 1981.

110. Ingberg, S.H., "Tests of the Severity of Building Fires," *National Fire Protection Association Q 22:1* (1928) 43–61.

111. Kawagoe, K., S. Takashi, andT. Sekine, "Estimation of Fire Temperature–Time Curve in Rooms," Occasional Report No. 11, Tokyo, Building Research Institute, 1963.

112. Law, M., "A Relationship Between Fire Grading and Building Design and Contents," Fire Research Note No. 877, Joint Fire Research Organisation, 1971.

113. Law, M., "Prediction of Fire Resistance," *Symposium No. 5, Fire Resistance Requirements for Buildings – a New Approach*, London Department of the Environment and Fire Office, Committee Joint Fire Research Organisation, HMSO, 1973.

114. Thomas, P., and A. Heselden, "Fully Developed Fires in Single Compartments – a Co-operative Research Programme of the Conseil International du Batiment," CIB Report No.20, Research Note No. 923, Borehamwood, England, Fire Research Station, 1972.

115. Law, M., "A Review of Formulae for T-Equivalent," *Fire Safety Science – Proceedings of the Fifth International Symposium,* International Association of Fire Safety Science, 1997.

116. Law, M., "Prediction of Fire Resistance," *Fire-Resistance Requirements for Buildings – A New Approach, Proceedings, Symposium No. 5,* Borehamwood, England, Fire Research Station, 1971.

117. Pettersson, O., S. Magnusson, and J. Thor, "Fire Engineering Design of Steel Structures," Swedish Institute of Steel Construction, 1976.

118. Magnusson, S., and S. Thelandersson, "Temperature–Time Curves of Complete Process of Fire Development," Civil Engineering and Building Construction Series No.65, Stockholm, ACTA Polytechnica Scandinavica, 1970.

119. Hannathy, T., and J. Mehaffey, "Post-Flashover Compartment Fires," *Fire and Materials* **7**:2 (1983) 49–61.

120. British Standards Institute, "Actions on Structures Exposed to Fire," Part 1.2: General Actions, *Eurocode 1: Actions on Structures,* BS EN 1991-1-2:2002.

121. Thomas, G., A. Buchanan, and C. Fleischmann, "Structural Fire Design: The Role of Time Equivalence," Fire Safety Science – Proceedings of the Fifth International Symposium, *International Association of Fire Safety Science,* 1997.

122. *Fire Engineering Design Guide,* Report of a Study Group of the New Zealand Structural Engineering Society and the New Zealand National Fire Protection Association, A. Buchanan, ed., 1994.

123. Kirby, B., et al., "Natural Fires in Large Scale Compartments," A British Steel Technical, Fire Research Station Collaborative Project, British Steel Technical Swinden Laboratories, 1994.